Innovation:
A Study of Technological Policy

Arthur Gerstenfeld, Ph.D. (M.I.T.)
Professor and Head of the Dept. of Management
Worcester Polytechnic Institute

Copyright © 1977 by

University Press of America™
division of
R.F. Publishing, Inc.
4710 Auth Place, S.E., Washington, D.C. 20023

All rights reserved

Printed in the United States of America

ISBN: 0-8191-0037-4

ALSO BY A. GERSTENFELD

Effective Management of Research and Development, (Addison-Wesley).

"Technological Forecasting," in Methods and Techniques of Business Forecasting, Edited by Butler, Kavesh, and Platt; (Prentice-Hall).

"A Study of Successful Projects, Unsuccessful Projects, and Projects in Process in West Germany," IEEE Transactions for Eng. Mgt.

"Marketing and R&D," Research Management.

"Technological Forecasting," Journal of Business.

"Engineering Turnover," Business Horizons.

"Decision Analysis for Increased Highway Safety," Sloan Management Review (with Paul Berger).

"Decision Analysis for Optimal Testing Sequence," Industrial Engineering (with Paul Berger).

CONTENTS

PREFACE vii

CHAPTER: Part One

1. THE PROCESS OF INNOVATION AND
   NATIONAL COMMITMENT.   1

   Overall Government Expenditure on
      R & D.   4
   R & D Direct Priorities.   9
   Electronic Data Processing   13
   Key Technologies Program   13
   Basic Research   15
   Industrial Joint Research
      Associations 20

2. STIMULATING INVENTION WITHIN THE
   ORGANIZATION   24

   Employed Inventors   24
   Application and Definition   27
   Compensation for Employees'
      Inventions 31
   Tax.   40
   Experience and Examples.   40

3. ASPECTS OF THE U.S. AND WEST GERMAN
   PATENT POLICIES.   46

   Growth of West German Patents in
      U.S. 46
   West German Broad Coverage vs. U.S.
      Specific 53
   Patent Life and Petty Patents.   57
   First Applicant Principle.   58
   Patent Importance and Diffusion
      Related to Firm Size 59

	Compulsory Licensing and Fees. . . .	64
	Euro-Patent.	70
4.	FUNDING FOR TECHNOLOGY: THE FIRM. .	75
	Centralization of Large Organizations	75
	Innovations from Medium and Small Sized Companies.	82
	Risk Financing and Venture Capital	86
	Private Financing.	86
	Corporate Support.	86
	European Economic Development Co. (EED).	87
	West German Venture Capital. . . .	88
5.	FACTORS RELATED TO SUCCESSFUL INNOVATIONS.	95
	Project Descriptions	95
	Demand Pull and Technology Push. . .	99
	Level of Effort and Early Warning Systems.	107
	Process vs. Product Innovations. . .	117
	Tax and Patent Considerations. . . .	120
6.	SOCIAL FORCES EFFECT UPON INNOVATION	124
	Third Generation Engineers and Scientists	124
	Motivating Forces for Projects in Work	126
	Example A.	130
	Example B.	134

7. TECHNOLOGICAL DIFFUSION AS PART OF
   THE INNOVATIVE PROCESS.   138

   Diffusion: The Firm's Technological
   System as an Information
   Processor   138
     a) Government Agencies and
        Laboratories   141
     b) Universities and Consultants . .   141
     c) Other Firms.   144
     d) Independent Inventors.   147
   Diffusion: Basic Research.   151
   Diffusion: University-Firm's
   Technological Interface   155

8. THE INFRASTRUCTURE NECESSARY FOR
   INNOVATION.   166

   Government.   168
   Industry.   169
   Legal Systems   172
   Financial Systems   174
   Education   175
   Environment and Diffusion   176

   Part Two (Case Studies)   178

1. The $100 Million Object Lesson. . . .   179
2. International Production Technology,
   Inc..   183
3. Innovation at Texas Instruments . . .   191

To

Jesse A. Gerstenfeld

PREFACE

This is a book for the reader who expects to be involved, one way or the other, with policies which influence technological innovations. Many of the readers may currently be public officials, managers, engineers, scientists, or technicians, engaged in technological innovations. Others may be persons who recognize the enormous importance of innovation in this complex world hence interested in improving their understanding of the innovative processes and the factors leading to success. Still another population for whom this book is directed are those students of management, engineering, and science who will be responsible for conducting and participating in projects that will lead to innovations in the years ahead.

The object of this book is to examine the innovative process and techniques for the better management of technological innovation, in the U.S., West Germany, and other countries. In the last few years there have been investigations from all corners of the world on trying to find a better understanding of what is involved in improving the management of technological innovation. There have been international and national seminars devoted to this subject, as well as a multitude of journal articles focussed on various thrusts for increasing our understanding of the innovative process. Universities offer complete courses - and parts of other

courses - in schools of management, engineering, and
public policy aimed at improving the policies concerned
with the management of innovation. The goals of
this book are to add new material to our increasing
knowledge of innovation while at the same time
synthesizing and organizing the concepts developed
by others in this field.

At the conclusion of each chapter a list of
suggested readings is included relating to the particular topic. It is suggested that those readers
using the book in the classroom might find value in
assigning various portions of the additional readings
in order to increase the perspective on a particular
subject. Similarly, the cases included in Part II of
the book are actual cases selected to generate discussion and thinking related to specific portions of
the text. Although much of the book was written
while I was in West Germany, and most of it is about
West Germany and the U.S., it is believed applicable
to other countries as well.

The book shall first consider the process of
innovation and national commitments focusing on the
role of government mechanisms. The next chapter
examines a unique German law which affects inventors,
namely the Employees' Inventions Law. Chapter Three
continues examining the Government's role and focuses
on aspects of the West German and U.S. patent policies.

Chapter Four concentrates on the firm level and
examines the role of funding for technology. The
following two chapters funnel down still further and
focus on specific technological projects. Chapter
Seven considers the general subject of diffusion, and
the final chapter synthesizes the findings and presents
a general perspective on innovation.

The data for Chapters 5 and 6 have been drawn
from both primary and secondary sources with emphasis on

primary sources. Published and unpublished data on present West German technological management were thoroughly searched.

The primary data were gathered through personal interviews with more than sixty managers representing industry, government, financial institutions, and academics.* The industrial research sector was the major focus of the study, but peripheral organizations that clearly affect technological development were included. The major industries included in the sample were the three top German industries as measured by R&D expenditures; namely, and in order of magnitude, chemical, electronics, and automotive. Interviews were with industrial research management, financial institutions, academics, legal, and government officials concerned with innovation.

While I used a carefully constructed and tested interview guide in the field research, the range of questions usually discussed during the interviews was more encompassing than that included in the formal interview guide. The flexibility of the oral interview method made it possible to concentrate in each case on practices that were most relevant to the study.

While some of the interviewees were more expert than others in different fields, I was interested in obtaining information on the range of subjects included under the general topic of innovation. It became my responsibility to attach increased validity to those responses that were closest to the individual fields of expertise.

---

*". . . Strange lands open their treasures more readily than the familiar world to the eye whose vision is bound by habit,. . ." Ralf Dahrendorf, <u>Society and Democracy in Germany</u>, preface to the German edition, New York: Anchor Books, 1965.

For example, the individuals most knowledgeable about public policy affecting innovation would be those people in Bonn employed by the Ministry for Research and Technology. However, the responses from leading research directors in industry would also be of some value in forming an accurate appraisal of the real system. In those cases where the person being interviewed did not feel that he was in a position to render a response, I encouraged him to pass so that the responses that I did receive were carefully thought out comments from people in a position to speak with validity on the particular subject. The responses were compared with the written material and in most cases checked with responses from other individuals in positions of knowledge.

Second interviews with the same individual were often more productive than the initial meeting because the rapport was already established. The pattern developed in which the interviewee would refer to some written documents, which I could obtain and study prior to the second interview. This second meeting permitted an easy exchange of ideas and a chance to concentrate on particular areas of interest. Sometimes the separation between interviews was as short as a week (just time to study some particular documents) and sometimes separated by several months, with correspondence taking place in between for the purpose of clarifying certain points.

During the course of the project, the available reports kept taking on increased importance. By careful studying of the written material, it no longer became necessary to waste interview time gathering information that was available. It, therefore, permitted me to use interview time to fill in the spaces left in the written material.

I tried to veer away from opinion, beliefs, and theories as much as possible, constantly probing for actual events, such as actual public policy and funding figures. The concept of learning through specific project analysis was emphasized by M.I.T. Professor Thomas Allen, for whom I have utmost respect and give great thanks.

It is my belief that many of the concepts discussed in this book are generalizable. The chapters which analyze 22 R&D projects have some findings consistent with other studies in other countries.

I presented my findings in West Germany and the United States to those individuals and groups who I felt were most knowledgeable about the system. It was not so much to see if there was agreement on the findings but mainly to be certain about the factual data.

A goal of this study is to stimulate and encourage the dialogue and debate which will lead toward more fruitful technological growth, as I am firmly convinced a major factor in economic and human survival is selective technological innovation. I hope that this study aids us in some small way toward a better understanding of the innovative process so that the necessary innovations in the future can be more effectively achieved.

While writing this book, I was fortunate to be offering a graduate seminar in Germany entitled Technology and Management. This seminar allowed me to try out many ideas with the students and to receive their feedback. Of course, the interviewees, who gave so freely of their time, deserve a special word of appreciation. I should like to list them all but will have to settle for a collective "thank you."

Special words of appreciation are being given to Dr. Bernd M. Kramer, Federal Ministry for Research and Technology, Bonn, Germany for his many helpful contributions. Equal appreciation is being acknowledged to Dr. Helmer Krupp, Director, Institute for Innovation, Karlsruhe, Germany. Our many hours of serious discussion served to crystalize and strengthen the many concepts within this study. Professor Otto Poensgen, University of Saarlandes, Saarbruken, Germany offered invaluable assistance through comments on my findings and during the seminar in which I presented these ideas at the University of Saarbruken. In the field of West German Patent Law and the West German Inventions' Law my appreciation is especially expressed to Dipl.-Ing. Heinz Frigger, Patent Attorney, Alfred Teves, Frankfurt.

The M.I.T. Center for Policy Alternatives deserves a thanks for supporting some of my expenses. Special thanks to Susan Gerstenfeld for encouragement and editing assistance.

PART I

CHAPTER I

The Process of Innovation and National Commitment

"Technological change is an important, if not the most important, factor responsible for economic growth."*

The impact of technology on economic growth has been recognized but perhaps not emphasized enough. As both individual enterprises and some entire nations are plagued by economic difficulties, there is not enough stress placed on the linkage between technology and economy.

The current energy crisis could really be considered a political and technological problem. The technology thrust is of enormous importance and Western countries are starting large scale technology programs toward finding alternate sources of energy.[1] Similarly, technology can aid in the solution of ecology problems. Whether one considers individual countries or companies, successful, well-directed innovation correlates with progress. Innovation is the bringing to market of a new product or process and the concept may or may not have originated in the innovating country or company. Innovation can be defined as a unit of technological change. An invention, if present, is part of the process of innovation. Research and Development (R&D) is often necessary to transform concepts to marketable products.

Recognizing the importance of innovation, I embarked upon a course of study which could examine

---

*E. Mansfield, <u>Technological Change</u>, New York: W.W. Norton and Company, 1971.

both national and individual enterprise technologies. The country and firms selected had to fulfill the requirement of a history of a strong economy and high technological growth.

West Germany fulfilled the criteria. It has an economy with more than a fifty percent contribution to the GNP from the goods producing sector, and there are some economists who say that this percentage is even increasing.

West Germany is the largest spender of funds for research and development after the United States.*

West Germany is also the largest foreign holder of U.S. patents.[2]

Mansfield addresses the subject of invention as follows:**

> Invention has been defined in many ways. According to one definition, an invention is a prescription for a new product or process that was not obvious to one skilled in the relevant art at the time the idea was generated. Other definitions add the requirement that the product or process must have prospective utility as well as novelty.

---

*OECD (Organization for Economic Cooperation and Development) "Patterns of Resources Devoted to Research and Experimental Development," Paris, May 1974. Gross expenditures for R&D place U.S. first, Germany second, and Japan, Third. (pg. iii).

**Mansfield, E., <u>The Economics of Technological Change</u>, Norton, 1968, p. 50.

This raises difficult questions as to how one is to find out whether a particular new product or process is prospectively useful, but it has the advantage of eliminating tinkering of an economically irrelevant sort. Thus, we include prospective utility as part of the definition.

Innovation refers to technology actually being used or applied for the first time. The R&D is necessary to develop the technology so that it can be used. The invention may take place anywhere along the R&D spectrum but prior to the innovation.

The innovation generally consists of:

. generation of an idea;

. problem solving; and

. implementation.

The generation of an idea contains elements of market needs and possible technology. The problem solving includes setting specific technical goals and designing alternative solutions. Implementation consists of the manufacturing engineering, tooling and market start-up.*

The distinctions between invention, R&D, and innovation are important.3 In this study we are concentrating on innovation. The processes of invention, R&D, and innovation are, of course, intimately intertwined. During the course of this book, I will describe many projects, some of which involve inventions, some of which do not.

---

*Utterback, James M., "Innovation in Industry and the Diffusion of Technology," Science, Vol. 183, Feb. 1974, pp. 620-26.

It is an illusion that in West Germany there is one clear policy with all participants having a clear role, with specified objectives, and appropriate strategies to reach each objective.[4] Rather, one sees the technological system as complex interactions involving lobbies, corporate interests, etc. These interactions lead to a policy that is largely one of reaction.

Overall Government Expenditure on R&D

A government's overall expenditures on R&D are a measure of commitment and are closely involved with innovation. The amount of funds spent by West Germany on research and development have been consistently increasing as shown in Exhibit 1-1.

During the course of this book there will be many comparisons made between the two largest spenders for R&D - the U.S. and West Germany.

The ratio between industries' contribution to R&D and the government contribution in West Germany is approximately 50-50. The German gross expenditure on R&D during the middle 1960's was approximately 1/15th of the United States. It is now close to 1/3. The industrial section of Exhibit 1-1 is the subject of Chapters 5 and 6, studied through an examination of 32 specific projects. The leading industrial spenders of R&D are the chemical, automotive, and electronics industries.

Exhibit 1-2 shows the comparison between West Germany and the United States R&D effort as a percentage of GNP and in terms of R&D manpower per 10,000 inhabitants. The exhibit shows that for the United States there has been a declining trend of the ratio

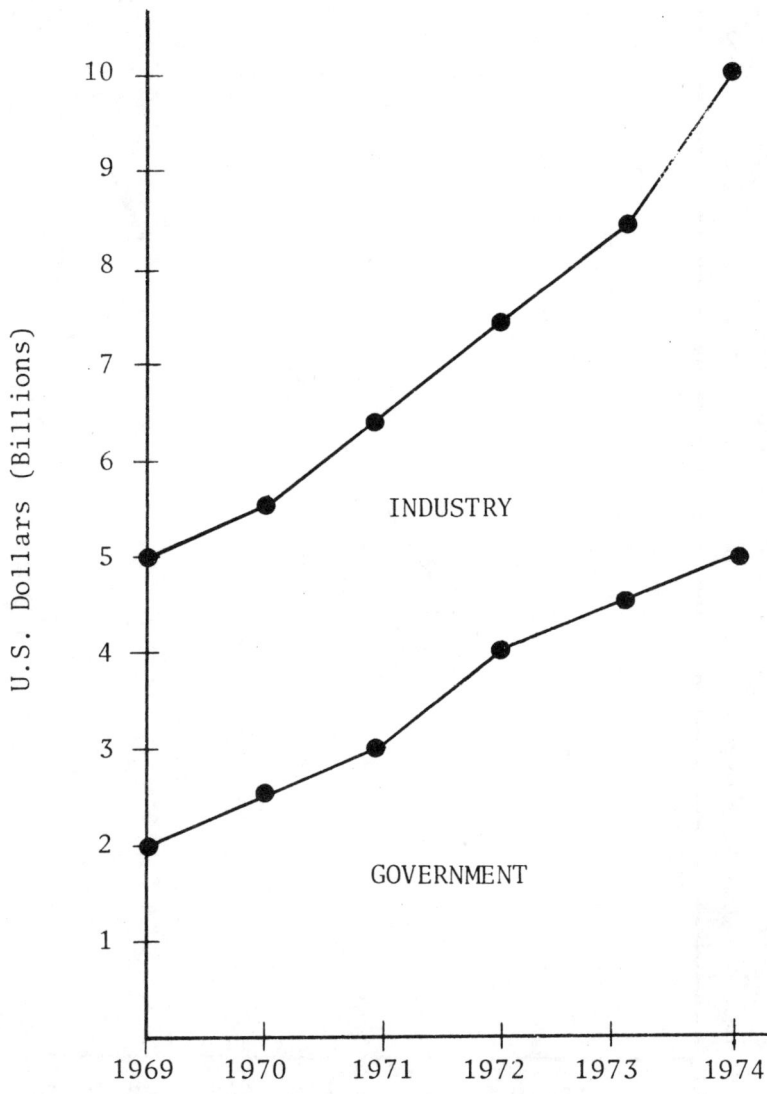

West German Expenditures for
Research and Development[5]
Exhibit 1-1

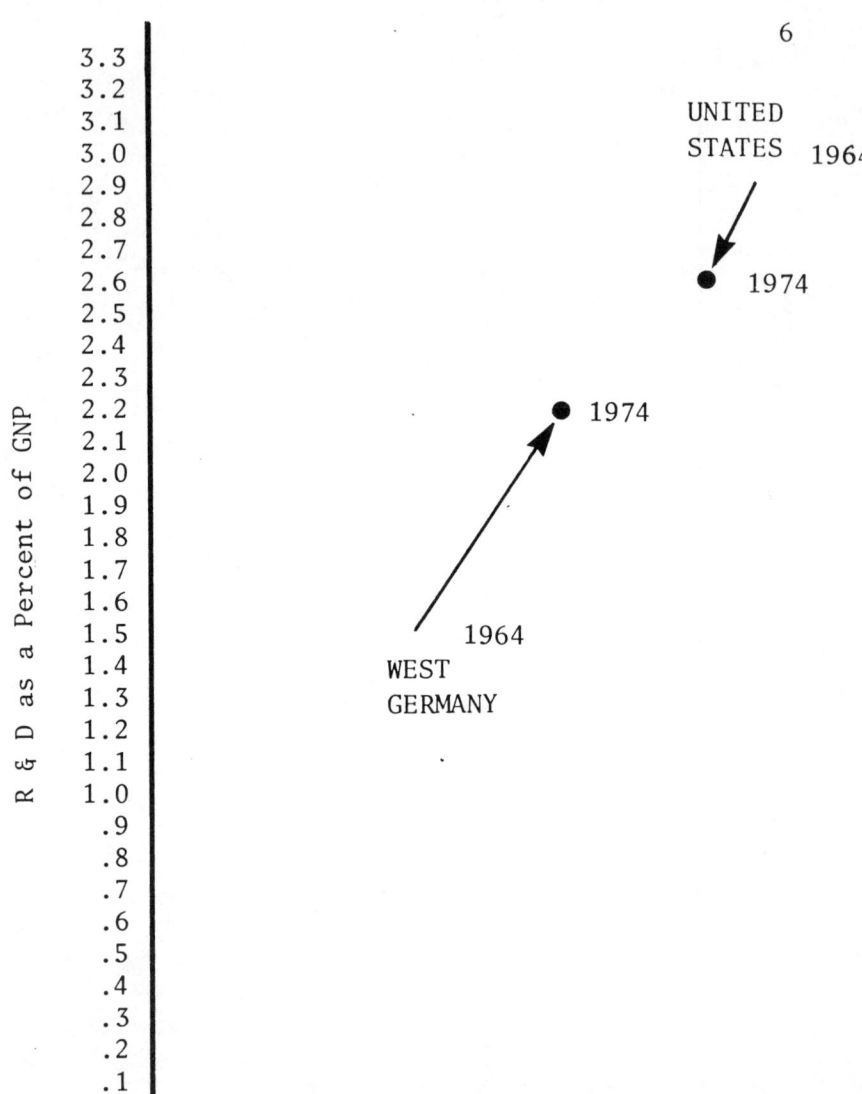

Changes in the Relative Amount of Resources Devoted to R & D Since 1963/1964 (OECD, 1974)[6]

Exhibit 1-2

of research and development to other areas of economic
activity. The United States is expected to devote
2.3% of its GNP to research and development in 1974,
down from 2.4% in 1973 and the high of 3.0% in 1964.

This decrease in the percentages for United
States R&D has been accompanied by a steady increase
by West Germany. The 1974 OECD reports stated the
following:

> The United States was the only country
> in which the percentage of GNP devoted to
> R&D actually fell during the period. . . .
> The growth was particularly marked in
> Germany where the relative manpower indica-
> tor also grew rapidly.[7]

This OECD statement should be tempered by
several facts. First, in terms of absolute amounts,
the United States is still spending approximately $32
billion as opposed to $10 billion for West Germany.[9]
Second, with the increased emphasis in the United
States on energy research and health, it is possible
that the U.S. trend will be reversed in the near
future, and it may spend more on R&D.

It is also likely that there will be a leveling
off of Germany's expenditures for R&D. I feel that it
is important to view these trends over a period of
years. The fallout from R&D expenditures occurs
several years later. It is for these reasons that
countries are often lulled into false economy decisions.

A marked change in the United States has been
underway in recent years in patterns of support.
Between the late fifties and late sixties, the federal
government in the United States supported over sixty
percent of the Nation's R&D investment. This is shown
in Exhibit 1-3. Since 1967 this support has been
dropping. At the same time, the German government support has been increasing. As mentioned previously, the

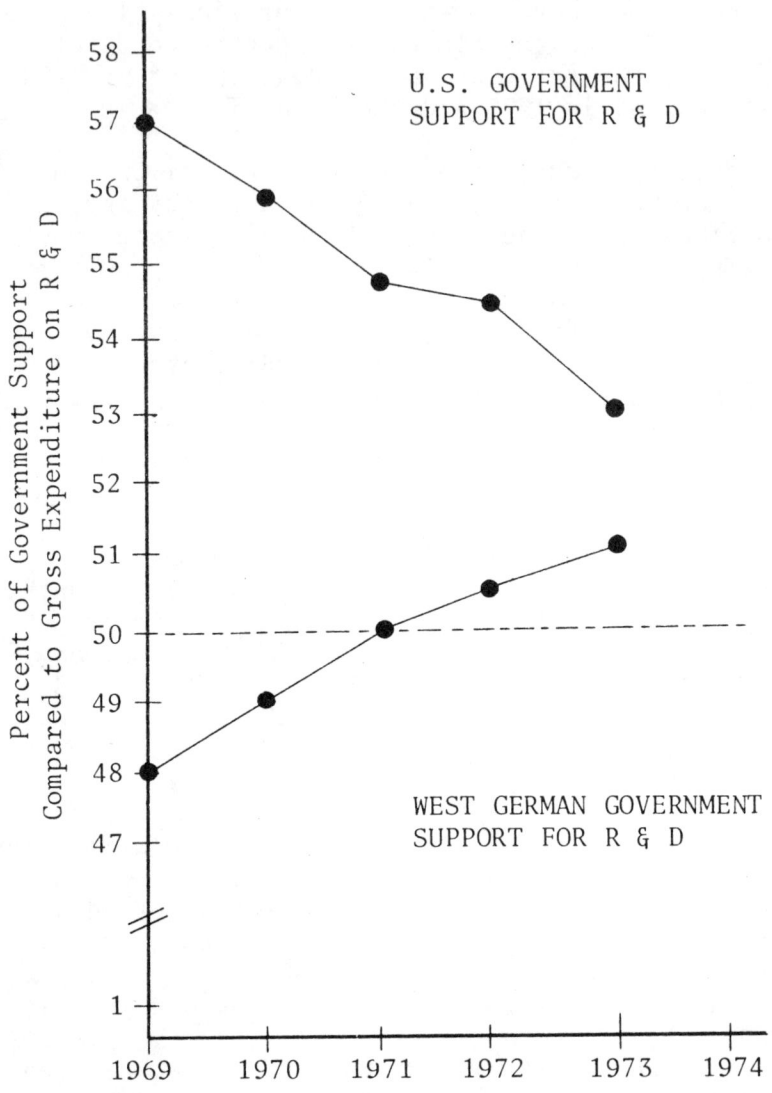

Government Support for R & D [8]

Exhibit 1-3

pressures are now building up within West Germany to level off on its support, so we might expect to see this fifty-one percent figure for some years in the future.

R&D Direct Priorities

The major responsibility for administering direct federal R&D programs in West Germany rests with the Ministry for Research and Technology.* (This includes all R&D budgets with the exception of the defense R&D expenditure and most basic science which will be discussed separately.)

Exhibit 1-4 shows the 1974 allocation of funds from the Ministry for Technology. The exhibit shows 34% of the current federal research to nuclear energy research. If one added to that the nuclear science program (8%) and the non-nuclear energy research (3%), we see a total effect of 45% going toward the energy problem. The current debate in Germany is not about the large R&D effort in the direction of energy but about the emphasis that the research is taking. There are some who believe that non-nuclear directions would be more fruitful, and, in the coming years, we will probably see more concentration in non-nuclear energy research.

In order to illustrate the small proportion of R&D funds being spent in Germany for defense, a comparison will be made with the U.S. Exhibit 1-5 compares U.S. and West German national security R&D for the year 1974. The United States spends 54% of its federal R&D

---

*Bernd Kramer, "West German R&D," Ministry for Research and Technology, Bonn, Germany; paper presented at George Washington University, Washington, D.C., September 1974.

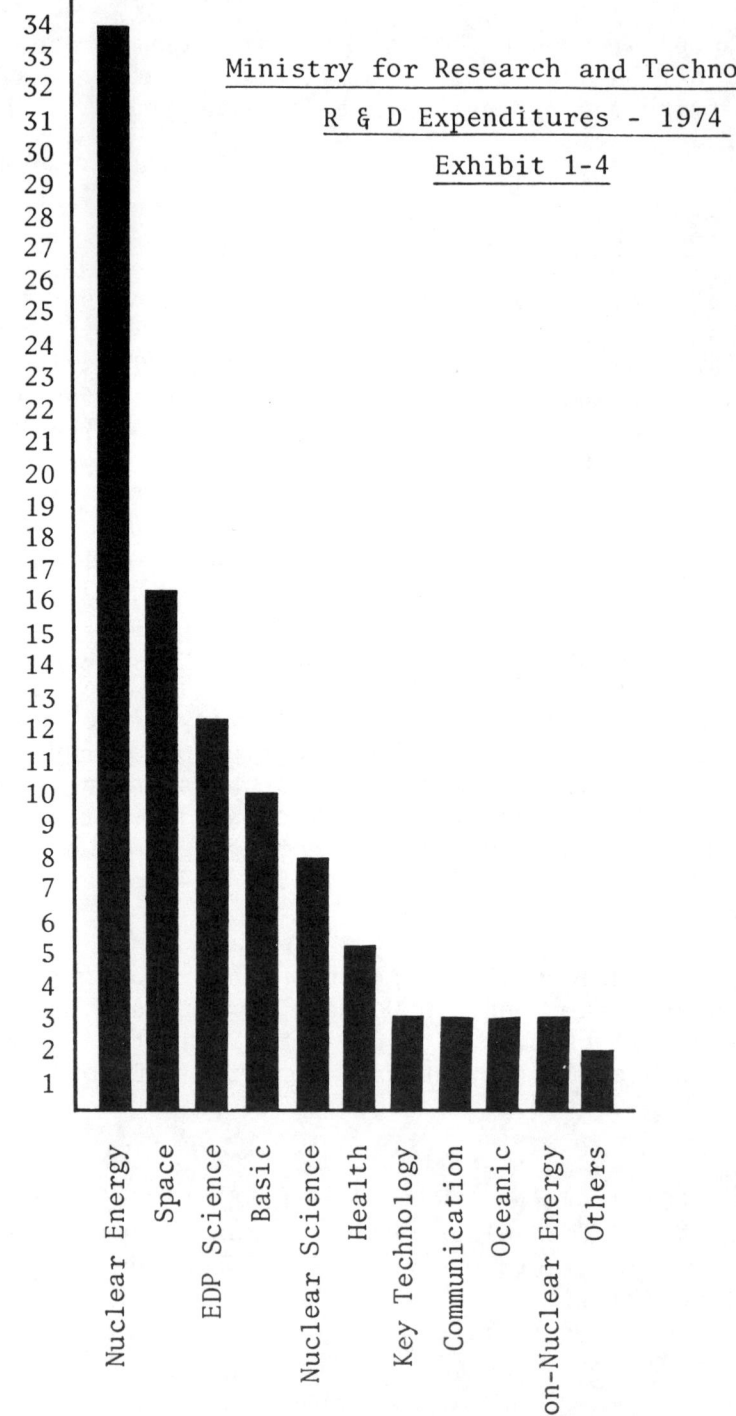

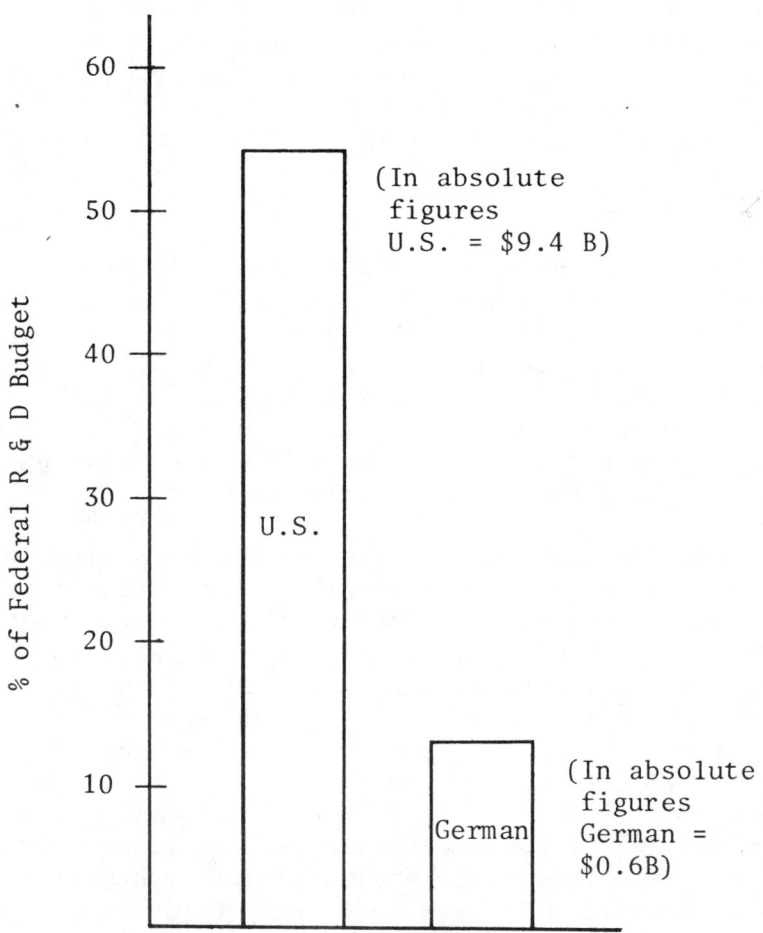

National Security R & D Expenditures[9]

Exhibit 1-5

funds on national security while West Germany spends 12.7% of its federal R&D funds for this function. In actual figures, the United States will spend $9.4 billion on national security while Germany will spend $0.6 billion. It has been argued by some that the large percentage of R&D spent by the United States on defense can be justified not only by the creation of national strength, but also by the stimulation of many spinoff technologies. However, it is highly debatable as to the cost/benefit value in the military spin-offs to civilian technology.[10]

It should be recognized that the electronic computer, numerical control, integrated circuits, atomic energy, and synthetic rubber are all significant inventions that have stemmed at least partly from military R&D. However, there can be little doubt that we could have obtained these benefits at lower cost and with greater certainty if comparable resources had been devoted directly to civilian purposes.

The classic example often cited for a civilian spinoff from a government effort is the teflon frying pan. It has been established that the plastic coated frying pan technology emanated from similar technologies in the aerospace field. However, if the country wanted to spent $1 billion for a new type of frying pan, it could obviously have performed a more directly focused R&D program.

Similar spinoffs occur in medical technology, but the circuitous routes to these new technologies result in enormously high cost-benefit ratios. It has been argued that with a long enough time delay, the spinoff technologies can be significant, but that remains to be proven.

We have, therefore, observed that while West Germany is putting emphasis on non-military research, the United States is still heavily directing its R&D effort toward defense. This is particularly important if it is coupled with the earlier established fact that the gap between the two countries on total R&D expenditures is decreasing.

Electronic Data Processing

The electronic data processing program in West Germany is a massive effort to establish a West German computer capability so as to decrease dependence on American firms. The government of West Germany has taken three steps to accomplish its goal of computer independence:

- o The government funds R&D in private firms to aid these companies in the development of new products.

- o The government has installed a "buy German computers" policy.

- o The government is restructuring the industrial sector in order to develop a few (or perhaps one) large, strong firm(s).

The government expenditures for promotion of EDP is shown in Exhibit 1-6. The budget showed an 18% increase for EDP over the previous year.

Key Technologies Program

Most of German industry finances its R&D actions by means of its own funds. (Exhibit 1-1 shows $5 billion being spent by German industry on R&D.) However, there are some technologies that the German government considers so vital that the government helps industry by sharing R&D risks with public funds. This "key technologies program" tries to stimulate industrial R&D in order to strengthen the economy.

EXHIBIT 1-6

Expenditures of the West German Research and Technology Ministry for the Promotion of the Electronic Data Processing Industry in 1974

	(in Million U.S. Dollars)
Federal Share for Research Program carried out in Universities	17.3
Special Research Projects carried out in Non-Profit Institutions	12.3
R&D of Future EDP Technologies	64.0
Investigations for new Applications of EDP	34.3
Society for Mathematics and Data Processing	16.1
Total	144.0

The amount of sharing varies with the risk of the project. For example, the research and development on a high speed magnetic train is being supported 100% by the government because of its high risk. For other projects the usual sharing arrangement is 50% of the R&D costs from the company and 50% from the government.

The optical industry has been a large recipient of this program (See Exhibit 1-7, Item d). The industry ran into intense competitive pressure from countries with lower labor costs. The optical industry could afford some R&D, but not enough to overcome the long-term crisis. The key technologies program has been sharing R&D costs in order to help this industry develop new products and processes.

Another example of the key technologies program is in the electronic instruments industry. This is shown in Exhibit 1-7, Item a. The program supports research on integrated switching techniques, opto-electronics and high frequency components.

Approximately $50 million has been spent on these programs for 1974, and some estimate that the 1975 expenditures will be close to $100 million.

Government Support of Basic Research
    in West Germany

Basic research as defined by the United States' National Science Foundation is original investigation for the advancement of scientific knowledge for which no specific commercial objectives exist.

Basic research is seldom a direct source of innovation. Basic research plays an important role in the production of knowledge. This research leads to education and thereby to innovation indirectly. Authors have pointed out that this often accounts for

EXHIBIT 1-7

Key Technologies Budget 1974

(Shared funding between West German Government and West German Industry to stimulate Innovation.)

		(Government share in Million U.S. Dollars)
a)	Electronic Instruments	22.8
b)	News and Information Techniques	7.2
c)	Physical Technologies	2.2
d)	Optics	12.0
e)	Humanization of the Workplace	3.6
f)	Production Improvements	.8
g)	General Innovation and Other Studies	1.0
	Total	49.6

the time lag between basic research and innovation.*

Studies have shown that applied research and development often stimulates basic research.** This relationship can be seen in Exhibit 1-8. The West German Fourth Science Report (1972) provides one example when it states that the development of the nuclear program raised problems for basic research. Questions arose during development necessitating basic investigations into the nature of matter, molecular structures, etc. Basic research can often precede applied research and development or basic research can be stimulated as a result of unanswered questions during applied research and development.

The Max Planck Society, composed of several institutes and successor of the Emperor Wilhelm Society (founded in 1911), holds the biggest potential for basic research in West Germany. The Society's achievements have earned it a high reputation both at home and abroad.

The Max Planck institutes vary considerably as to size and personnel; some employ not more than twelve scientists, others have a staff of over a thousand. A director's term of office is restricted to an initial period of seven years which may be prolonged.

With a view to achieving greater flexibility of research programs, the departmental structure as shown in Exhibit 1-9 was introduced in the sixties.

At present the Max Planck Society maintains forty-seven research institutes of which thirty-eight

---

*Allen, T.J., Ph.D. thesis, Mass. Inst. of Technology (1966).

**C.W. Sherwin and R.S. Isenson, Science, (1967), pp. 156-57.

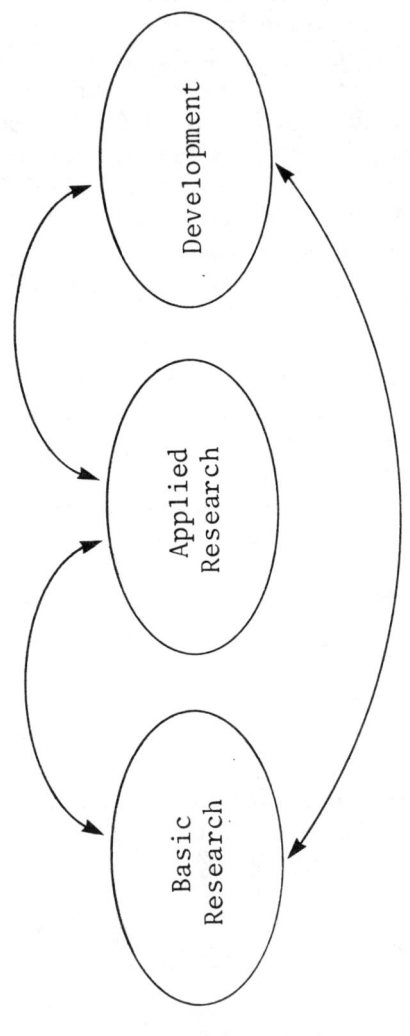

Basic-Applied-Development Loops

Exhibit 1-8

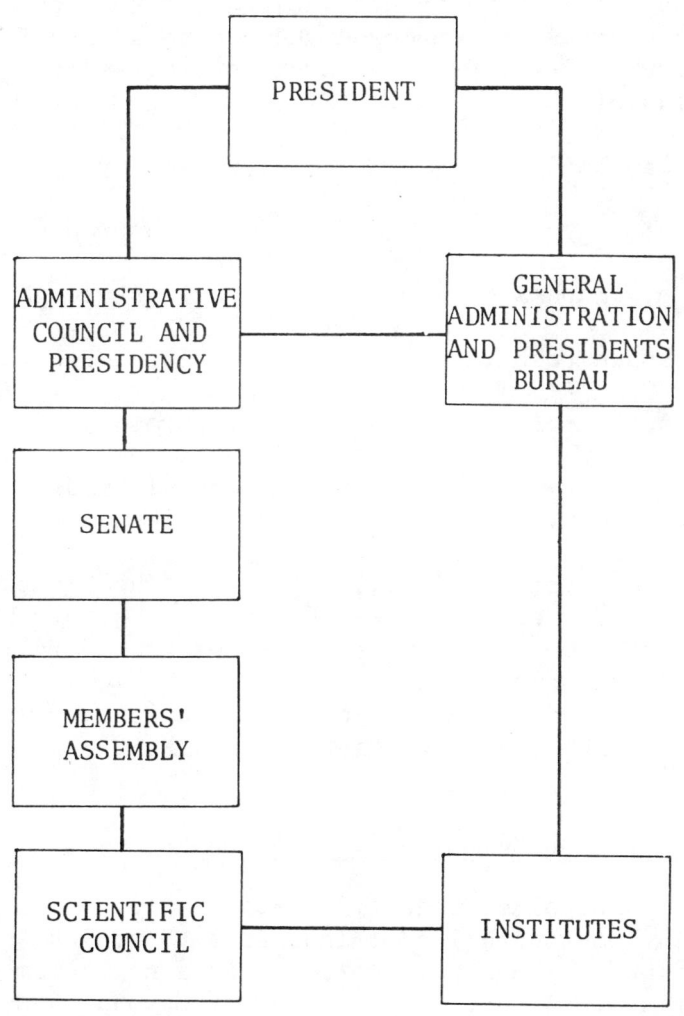

Max Planck Society for
the Advancement of Science
Exhibit 1-9

carry out research in natural sciences. More than 4,000 scientists are employed and the total staff exceeds 8,000. Priority is being given to research in the fields of:

- biochemistry
- biology
- plasmaphysics
- space research
- nuclear

- astronomy
- radioastronomy
- solid state physics
- meteorology
- futurology
- social & educational sciences

There exists a close link between the Planck Society and university research. The funds for the Society, that come from public sources (90%) are shared by the federal and the state governments evenly. The rest of the funds take the form of donations from industry and income from commissioned work.

Industrial Joint Research Associations

In 1958, eight industrial research associations formed the Confederation of Industrial Research Associations. It has grown to its present 78 member associations. Almost all branches of German industry are represented. It is an autonomous organization which follows the broad objective of supporting and financing co-operative research. The Industrial Research Associations' main financial burden is carried by industry's own contributions. In addition, it receives some public funds.

The goals of the Industrial Research Association are to:

- finance research projects conceived by the member associations

- to coordinate projects

- to facilitate the exchange of experience between its members

- to act as a link between members and the public administration.

The member associations vary in size and structure and the majority have their own research institutes. Some maintain institutes at technical universities, and some sub-contract research to universities or to independent research institutes. Exhibit 1-10 shows the relationship between government and industry for the industrial joint research associations.

These associations receive funds from the government as well as contribute their own funds to do long-run research which the individual corporations could not afford. For example, the automobile association might be working on a low pollution motor, a different braking system, carburetion etc. In this case, the five German automobile manufacturers would all participate and share the findings. The system works fine as long as the research is somewhat theoretical and long-run. As solutions become closer, the competition then develops among manufacturers and some difficulties arise.

The West German government has a major commitment to research. The government support takes many forms, such as 1) direct spending as in energy, medicine, etc.; 2) shared spending, as in the key technologies program; 3) funding of basic research, as in university and Max Planck support; and 4) in the contributions to the industrial joint research associations.

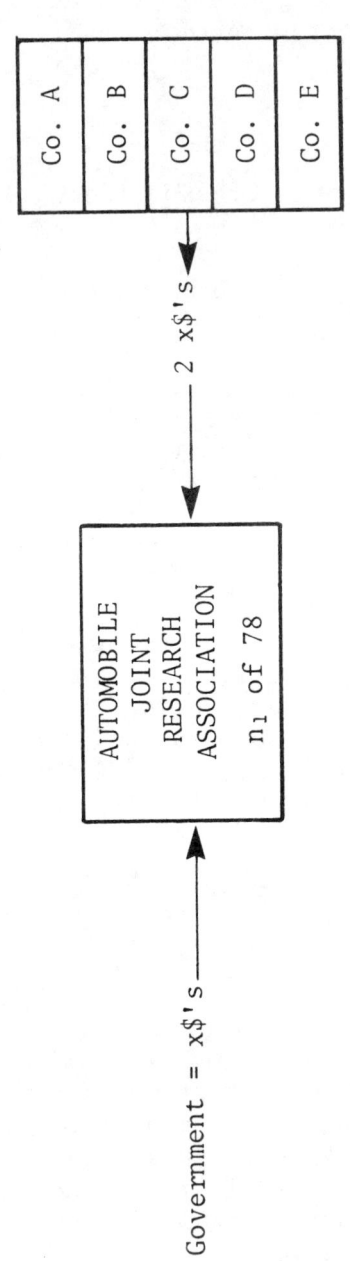

Industrial Joint Research Association

Exhibit 1-10

Chapter 1

Selected Readings

Haeffner, Erik A., "The Innovation Process," Technology Review, M.I.T. Cambridge, Mass., March/April 1973.

Mansfield, E., Technological Change, New York: W. W. Norton & Co., 1971.

Mansfield, E. et al., Research and Development in the Modern Corporation, New York: Macmillan, 1971.

Marquis, D., "The Anatomy of Successful Innovation," Innovation, Number Seven, 1969.

Stead, H., "The Costs of Technological Innovation," Research Policy, 5, 1976.

Steele, L., Innovation in Big Business, New York: Elsevier, 1975.

Utterback, James M., "Innovation in Industry and the Diffusion of Technology," Science, Vol. 183, Feb. 1974.

Wiesner, Jerome B., "Technology Is For Mankind," Technology Review, M.I.T. Cambridge, Mass., May 1973.

CHAPTER 2

Stimulating Invention Within the Organization*

COMPENSATING EMPLOYED INVENTORS CONTINUES TO PLAGUE GOVERNMENT, COMPANIES, ENGINEERS.

New York - An industry-wide controversy over the system of compensating employed inventors for their work is brewing while legislative proposals continue to simmer in Congress.

Engineers repeatedly charge mistreatment of their personal interests by companies that require the signing of stringent employer-employee invention agreements as a condition of employment.**

Employed Inventors

The above quotation explains the dilemma that exists in trying to create incentives for innovation. The patent law was originally devised as just such an incentive. However, in many respects the patent laws share the fate of many other laws that regulate legal relationships in the economy: economic and industrial progress has marched on and passed them by.[1]

In the United States, most employed inventors sign over all rights to the employer and therefore

---

*Portions of this chapter were presented by Gerstenfeld at the Seminar on Technological Innovation, Bonn, Germany, April 1976.

**Electronic Engineering Times, CMP Publications, Great Neck, New York, May 24, 1976.

receive only psychic rewards and sometimes $1 as a token of appreciation.2 The incentive for innovation in terms of patent rewards for the individual has largely been lost. The number of patents being applied for and issued is decreasing in the United States.3 Whether this is directly attributed to the loss of incentive or to other factors is, of course, hard to determine. However, it is clear that with the vast number of inventions coming from employed inventors, it is time we reexamine the system.

I present here the German Law on Inventions since it appears as a vehicle whereby the employed inventor can receive extra rewards for his efforts.4 The original intent of the Employees' Inventions Law was to set up incentives to encourage invention by employed inventors. One obviously realizes the intrinsic rewards that inventors receive, but, realistically the extrinsic rewards serve as important stimuli.

Under the classic patent laws in the Western world, the right to a patent belongs to the inventor.5 Under the basic principles of labor law, the fruit of the labor belongs to the employer. We therefore have a situation as shown in Exhibit 2-1.

In the United States the decision is clearly made in favor of the employer. This is done at the expense of the inventor. The inventor receives little or no extrinsic rewards for invention. Groups in the United States have been writing about the seriousness of this problem and have been proposing an inventors day in which one day be set aside each year to honor inventors. Other techniques such as medals have also been suggested.6

These incentives only show that the United States is starting to become aware of the losses that the nation faces by not motivating invention. But the solutions are far from adequate. I present the German Employees' Inventions Law as an illustration of one

26

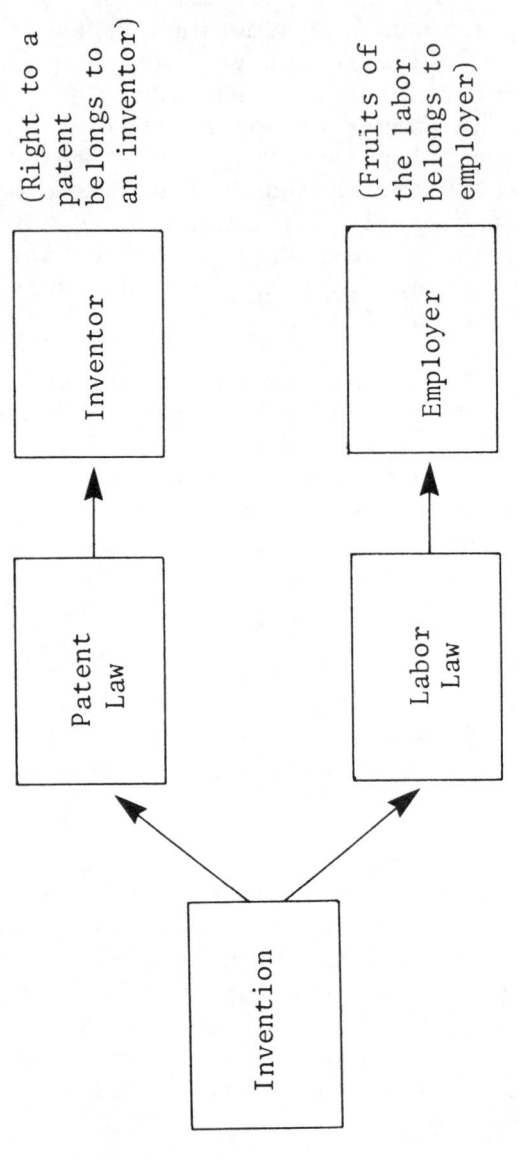

The Patent Law - Labor Law Conflict: In General

Exhibit 2-1

country's technique for encouraging invention.[7]

Application and Definition

This Law applies to inventions and to technical improvement proposals in West Germany made by employees in private employment, by employees in public service, by civil servants, and by members of the armed forces. This Law covers patents, petty patents and what are called qualified proposals for a technical improvement. The German Employees' Inventions Law states further that other proposals not ranking among those qualified for a technical improvement are subject to collective agreement or agreements concluded between employer and works council.

Inventions within the meaning of this Law may be either service or free. Service inventions are those which result from the employee's task within the organization or are dominantly owing to the organization's experience or activity.

Other inventions of an employee shall be free inventions. They are, however, subject to the following restriction: the inventor must notify his employer of his invention to enable the employer to check whether a free invention is indeed concerned.

This obligation to notify does not exist if it is obvious that the invention cannot be used within the employer's range of activities.

With respect to a free invention, the inventor must offer it to his employer; in other words, the inventor must offer his employer a non-exclusive license to use the invention on reasonable terms, if the invention falls within the range of the activities of the employer's enterprise.

Any employee making a service invention must

report the invention to his employer immediately in a special written notice. The employer shall then inform his employee without delay, and, in writing, of the date the report was received. In the report, the employee must describe the technical problem, its solution, and how he arrived at the service invention. Any existing notes necessary for an understanding of the invention shall be attached. The report shall include the extent of co-workers' contributions, and the report should underline the contribution which the employee making the report considers to be his own.

A service invention shall become free under these conditions:

(a) if the employer releases it by a written statement; or

(b) if the employer makes only a "limited" claim to it. "Limited" as used in this sense means that the invention is free, but on the other hand the employer has a non-exclusive right to use it; or

(c) if the employer has not made a claim to it within four months after receiving a proper invention report.

In my interviews with research managers and patent attorneys, I was told that this third provision is of major concern to the organization. For, in essence, this states that if the organization does not keep alert and allows the four months to go by without a response, the service invention becomes a free invention and is the property of the inventor.

After having applied for and having obtained a patent on his own behalf at the Patent Office, the employed inventor may license it to any other company, including a competitor, and receive a royalty rate as shown in Exhibit 2-2. The Law is quite clearly in

29

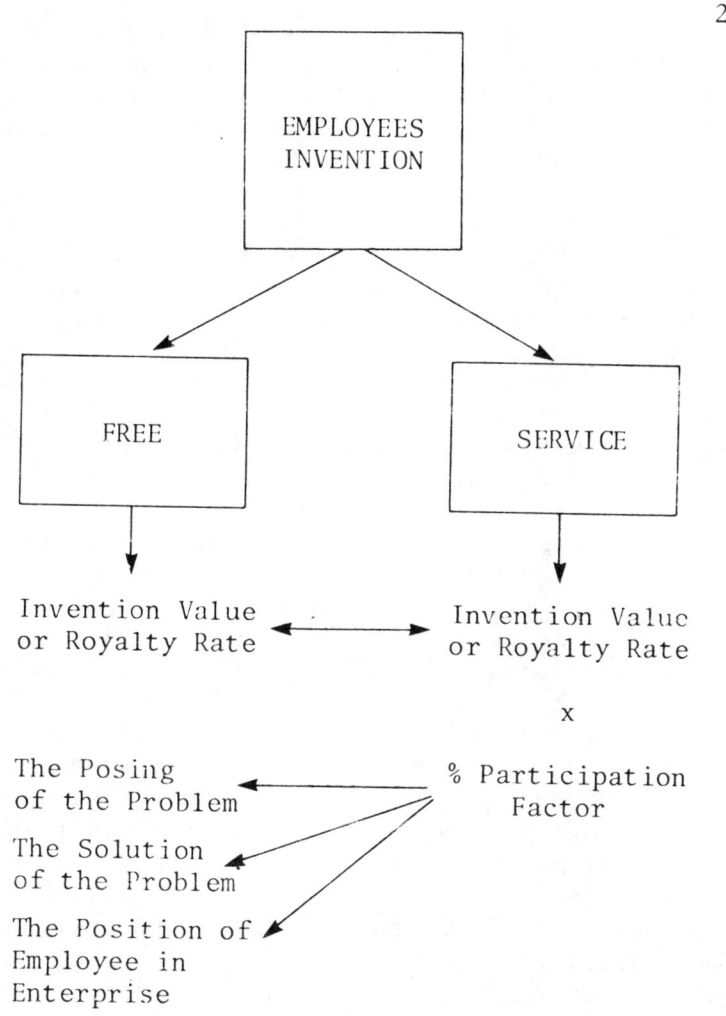

West German
Employees Inventions with Participation Factor
Exhibit 2-2

favor of the inventor and is designed to provide an incentive for innovation. The companies, in general, however, feel that although the Law is burdensome to management, the burden is worth the stimulus it provides to the employees.

For service inventions, the nature and amount of compensation shall be established by agreement between the employer and the employee within a reasonable time after the claim to a service invention. If two or more employees have contributed to a service invention, compensation shall be determined separately for each of them. The employer must notify the employees of the total amount of compensation awarded and of the share assigned to each inventor.

The provisions of the Law may not be modified by contract to the detriment of the employee prior to the notification of the service invention. <u>This means, for example, that an employment contract may not provide that all inventions pass to the employer.</u> There cannot be a contract that states that compensation for inventions is included in the generous salary paid to the employee. Under certain circumstances, an employer must maintain secrecy concerning an employee's invention that has been reported to him; otherwise, the employer will be liable for damages. An employee must keep a service invention secret as long as it has not become free. In all disputes between employer and employee arising as a result of this Law, petition may be made at any time to the Arbitration Board. The Arbitration Board shall seek an amicable settlement.

Inventions made by employees in offices of the Federal government and state governments shall generally be governed by the provisions relating to employees in private employment. Inventions made by professors, lecturers, and scientific assistants shall be free inventions.

Compensation for Employees' Inventions

In accordance with the spirit of the Law, the following, in particular, are significant in assessing compensation: <u>the commercial applicability</u> of the service invention, <u>the duties and position of the employee</u> within the enterprise, and <u>the enterprise's contribution</u> to the invention concerned. As a rule, therefore, in order to ascertain the compensation, the commercial applicability of the invention will first be determined. With inventions either arising from an employee's duties in his enterprise or substantially based upon experience or activities there, (service), a deduction by the company is made which will take into account the duties and position of the employee in his enterprise as well as the enterprise's contribution to the invention. The employee's share in the invention value, arrived at after the deduction, is expressed as the participation factor, given as a percentage.

As a rule, the invention value assigned to inventions created in the enterprise can be determined by three different methods:

a) <u>Establishing the invention value by a license analogy</u>

> Under this method, the invention value is ascertained on the basis of the royalty rate customarily applied to comparable free inventions, as shown in Exhibit 2-2.

b) <u>Establishing the invention value by reference to a measurable benefit to the enterprise</u>

> The invention value can also be determined by considering the measurable benefit obtained by

the enterprise from the use of the invention.

c) <u>Estimation of the invention value</u>

The invention value can also be estimated.

Establishing the invention value according to the measurable benefit to the enterprise occurs primarily with inventions that produce savings, as well as with inventions directed to improvements. The method for establishing invention value by reference to the measurable benefit to the enterprise has the disadvantage that it is often difficult to determine the benefit. In many cases, it may, indeed, be possible to calculate the benefit resulting from the achieved price reduction of the starting material, the reduced costs for wages, electricity, or repairs, or from an increased yield. An estimation of the invention value may only be applied if it is impossible, or possible only at a disproportionately high cost, to determine the value using the other methods.

The following general range of percentages based on sales volume is intended as a reference in determining a royalty rate suitable for a specific branch of industry:

- for the electrical industry, a royalty rate of   1/2 to 5%

- for the machine and tool industry, a royalty rate of 1/3 to 10%

- for the chemical industry, a royalty rate of   2 to 5%

- for pharmaceutical fields, a royalty rate of   2 to 10%

In the case of an extremely large sales volume--that is, one in excess of $400,000--, a reduction in the royalty rate, in accord with general practice, may be applied on the basis of an inverse scale. See Appendix A.

For all service inventions, the invention value must be reduced by a given amount to allow for the fact that it is not a free invention. This can be seen in Exhibit 2-2. The employee's share in the invention value after such a deduction is made is expressed as the participation factor, given as a percentage.

Again, the participation factor depends upon:

(a) points assigned in relation to the posing of the problem;

(b) points assigned in relation to the solution of the problem;

(c) points assigned in relation to the duties and position in the enterprise.

Exhibit 2-3 shows the relationship between the number of points and the participation factor given as percentages of the royalty rate. It can be seen in Exhibit 2-3 that the inventor who receives a low number of points receives a low participation factor, thus a low percentage of the royalties. On the other hand, the inventor who receives a high number of points receives a high percentage of the royalties. The ultimate number of points is 20 which corresponds to 100% participation, or essentially a free invention.

The sum obtained from adding (a), (b), and (c) need not be a whole number. If intermediate values were taken as points of reference, the participation factor will be somewhere in between the figures mentioned. The values of 20 and 100 have been placed in parentheses since, in this case, at least, a free

Number of Points a+b+c	Participation Factor as a Percentage (A)	
3	2%	.Minimum participation .Essentially the invention is company property .Royalties to corporation (98%)
4	4%	
5	7%	
6	10%	
7	13%	
8	15%	
9	18%	
10	21%	
11	25%	
12	32%	
13	39%	
14	47%	
15	55%	
16	63%	
17	72%	
18	81%	
19	90%	
(20)	(100%)	.Full participation .Essentially a free invention .Full royalties to inventor

WEST GERMAN

Employees' Inventions Participation Factor

Exhibit 2-3

invention is involved.

The employee's share in making a service invention increases in proportion to the greater initiative on his part in posing the problem and in his contribution in recognizing the need. These factors are calculated on the basis of the following:

1) the employer posed a problem for him indicating directly the approach to be taken in solving it (1 point);

2) the employer posed a problem for him without indicating directly the approach to be taken in solving it (2 points);

3) as a result of the employee's knowledge--obtained through his employment--of the company's needs, but the inventor did not recognize these needs himself (3 points);

4) as a result of the employee's knowledge--obtained through his employment--of the company's needs, and the inventor recognized these needs himself (4 points);

5) the employee posed a problem falling within his range of duties (5 points);

6) the employee posed a problem falling outside his range of duties (6 points).

The following aspects are considered in arriving at the points assigned to solving a problem:

1) the employer aided the inventor by providing technical assistance (2 points);

2) the solution found was based on activity or knowledge customary in the enterprise (4 points);

3) the solution was found with the aid of the inventor's professional approach to the problem (6 points).

If all of these characteristics apply, i.e., the employer aided the inventor with technical assistance, etc., only 1 point is assigned. If, on the other hand, the solution was found with none of these features present, the maximum of 6 points is assigned.

An employee's share of the invention value diminishes proportionally as his position provides him with more insight into production and development activities in his enterprise, and as the expectation increases that he will contribute toward technical achievements of his employer because of the position and salary paid to him at the time the invention report was made. Distinctions might be made among the following categories so that the assigned number of points increases in proportion to a diminished expectation of performance level:

Group 1: placed first are the heads of all research departments within a given company and the technical directors of larger corporations (1 point);

Group 2: to this group belong the heads or managers of development sections or departments, as well as section managers in research (2 points);

Group 3: to this group belong the supervisors of a complete production unit or research engineers and chemists (3 points);

Group 4: included here are those supervising production and those engineers and chemists employed in development work (4 points);

Group 5: to this group belong employees who have received a higher scientific education at a university, a polytechnic institute, a technical institute of higher learning, or an engineering school or similar institution and whose work is related to production. Such employees can be expected to possess a keen technical interest in addition to the ability to solve certain problems of construction or methods (5 points);

Group 6: members of this group carry out supervision at a lower level (e.g., master craftsmen, shop masters or foremen) or have received a somewhat better-grounded technically oriented education (e.g., as laboratory chemists or technical assistants) (6 points);

Group 7: this group includes employees with a practical technical training (e.g., skilled workers, laboratory technicians, assembly men, draftsmen) even if they have already been assigned minor supervisory duties (e.g., as foremen, substituting foremen, shift bosses, group leaders) (7 points);

Group 8: to this group belong employees for the most part without training for the activities to be performed in the enterprise (e.g., untrained workmen, temporary (unskilled) help, trainees, apprentices) (8 points).

The preceding arrangement of categories can only serve as a point of reference. Classification into a particular group must always consider actual circumstances. In smaller companies, for example, the heads of research departments will frequently not belong in Group 1, but--according to specific circumstances--to groups 2, 3, or 4. Furthermore, classification of activities into production, development, or research is not always justified, since in some companies, for instance, employees working in development are more closely associated with possibilities for inventions than are employees actively engaged in research.

Compensation calculated on the basis of the invention value and the participation factor can be expressed in terms of the following equation:

$$V = E \times A$$

where:

V = the compensation payable;

E = the invention value;

A = the participation factor given as a percentage.

The invention value established by the license analogy is given as:

$$E = B \times L$$

where:

E = the invention value;

B = the unit of reference;

L = the royalty rate given as a percentage.

In the second equation, the unit of reference may be a sum of money or a number of pieces. If the unit of reference is a sum of money, the royalty rate is given as a percentage (e.g., 3% of $50,000). On the other hand, if a number of pieces or a unit of weight is taken as the unit of reference, the royalty rate amounts to a sum of money per piece or per weight unit (e.g., $0.10 per piece or weight unit of the product sold).

Taken together, the above equation yields the following for establishing compensation payments by the license analogy method:

$$V = B \times L \times A$$

For the above, B always denotes the corresponding unit of reference (sales volume, production). B may cover a specific cycle in time (e.g., one year). The equation thus yields compensation for the entire term (V) or for a specific period of time (hereinafter referred to as V_j for a yearly assessment). If the compensation amount was established by combining the license analogy method with the sales volume (U), calculation would be made in accordance with this equation:

$$V = U \times L \times A$$

or for a yearly assessment:

$$V_j = U_j \times L \times A$$

Example: Given annual sales of $350,000, a royalty rate of 5%, and a participation factor of (a + b + c = 11) 25%; compensation for a single year would be $4,375:

$$V_j = 350,000 \times .05 \times .25$$

$$V_j = \$4,375 \text{ (Inventor's yearly income over salary for that invention)}$$

Tax

For all compensation received under this law, the individual income tax is reduced by one half the usual tax paid on income earned in other ways - still another incentive for innovation.

Experience and Examples

After this description of the Inventors' Law, I shall now discuss some aspects of its practical application and its effects. The Law must be observed. Of course, the employee still needs to know his rights and must also be willing to take advantage of them. This is still not always the case, as will be seen toward the end of this chapter. It is, above all, the large enterprises which adhere to the Law, frequently using forms to cover certain aspects of its administration; for example, forms for reporting an invention.

I should like to provide a more detailed discussion of service inventions, particularly where there are several contributing inventors, as that is becoming more the rule than the exception. A service invention can only be one which is associated in a specific way with the employee's enterprise. As mentioned earlier, under German law, a service invention must either result from the employee's tasks or be essentially based upon experience or activities in the enterprise.

No difficulties are caused by inventions evolving from research and development activities of engineers, chemists, etc. Generally, it will be quite evident that these persons were carrying out their duties at the time the invention was made, or that the experience or activities in the enterprise had a decisive influence on the making of the invention. It is rather the borderline cases which occasionally present difficulties. In one case, an opinion was requested concerning whether a salesman entrusted with market research, whose study of competitive products led him to contribute an important aspect of an invention, was to be treated as the author of a service invention or as an independent inventor. A similar situation arose in the case of a newly employed business manager who contributed a significant idea toward redesigning a machine of utmost importance to that business. The Arbitration Board has also frequently had to consider the question of whether or not the contributions made by lawyers come under the category of a service invention or of a free invention. In these particular cases, the Arbitration Board suggested that the inventions were service inventions.

These are borderline cases that may occur anywhere. The decision as to whether his contribution is or is not to be considered a service invention is not left to the employee, since he is also obliged to report to his employer inventions which he feels to be free inventions.

Uncertainty and vagueness are more frequently encountered in determining who is to be considered a co-inventor, particularly where the contributions of individual members of a work-team taken separately are insufficient to support a patent application, but where the finished product of such teamwork is clearly patentable.[8]

In summary, the Law requires an invention to be claimed by the company, in writing, within a period of four months, otherwise it becomes free. The employee may then dispose of it as he wishes--he may even sell it to a competing firm. The employees inventions law operates well in large enterprises. Firms of a medium or smaller size frequently do not act in accordance with the rules.

As already indicated, the pivot point of the Law is the obligation to compensate the inventor. Its purpose is to provide the employed inventor with a fair return based on the advantage to the employer of being able to obtain an industrial property right. The legislature has assumed that an invention is not the normal result of work. Phrased differently, an employee cannot commit himself to devise inventions, because this would be contrary to the essence of an invention.

In most instances compensation is assessed according to a method employing a license analogy: one determines the type of license contract that might be concluded in similar circumstances.

Of course, an employed inventor will not receive the same amount of compensation as a "free" inventor: he is employed by the enterprise, is working with materials provided by the enterprise and does not participate in the costs and risks during development. He therefore receives only a portion of the value assigned to an invention.

Some may be skeptical whether a point system can result in reasonable compensation. Experience plays an important part. Nevertheless, this type of calculation is one which engineers and chemists follow with ease, and it lends itself to a more uniform application of compensation rates.

The interviews and the data from the Arbitration Board show the value arrived at for the participation factor in most cases lies between 10 percent and 32 percent (6 points to 12 points) of the amount which would be paid to a free inventor.

It is not always easy, and experience is necessary, to arrive at a royalty rate that would be customary for comparable license agreements; for example, about three to five percent of the value of a novel machine, two to three percent of the value of an automatic flash mechanism for a camera, five percent of the sales for a particular drug, or only 0.3 percent of the sales of a radio.

There are instances in which it is hardly possible to base the analogy on production or sales, because the invention results in savings within the firm, either reducing the manufacturing costs of the process or product or having other advantages; for example, an invention in a large chemical company which considerably improved the utilization of exothermic heat developed during a chemical process, thus contributing a significant economic factor. In another instance, an invention greatly reduced the waste encountered in the manufacture of transistors.

At the suggestion of industry, the Arbitration Board has taken, as a guide for invention values, figures ranging from one-third to one-eighth of the actual savings realized. No enterprise pays to a free inventor an amount equal to the savings it hopes to make if his invention is utilized. The enterprise has to take a certain risk as to whether the invention will prove itself; it also has to take investments and overhead costs into account. Furthermore, the enterprise is interested in making a profit in acquiring that invention. In the chemical case referred to above, which concerned the yield of nitrogen, the rate arrived at amounted to 20 percent of the profit estimated over ten years. This amount was reduced by the participation factor. For an estimated profit of over $400,000, the compensation paid to the inventor amounted to about $32,000.

A survey of the fifteen years during which the Law on Employees' Inventions has been in force in the Federal Republic of Germany, shows that its effects have been beneficial. The Law has not prevailed in

all respects. Many inventors are not aware of the Law
or dare not take advantage of their rights under it.
It is true that for inventions that are used, hundreds
or thousands of dollars are frequently being paid; for
successful inventions these amounts may be $50,000 and
even more. Nevertheless, this is no burden to industry
compared with amounts spent for social security contributions and taxes, in addition to wages and salaries. Federal statistics indicate that a reduction of
taxes amounting to $5 million was estimated for a year
as a result of the 50% tax exemption allowed for compensation paid to inventors. The Government thus
contributes to the promotion of inventive activity
through such tax allowances. Not only the economy as
a whole, but also the employer benefits from such
incentive, since it augments the employee's interest
in his work.

Selected Readings

Allen, T. J., A. Gerstenfeld, and P. Gerstberger "The Problem of Internal Consulting in Research and Development Organizations," M.I.T. Working Paper, July 1968.

Etzioni, A., "An Engineer - Social Science Team at Work," Technology Review, Jan. 1975.

Journal of the Association for the Advancement of Invention and Innovation, Arlington, Virginia, Jan.-Feb. 1974.

Neumeyer, F., The Employed Inventor in the United States, Cambridge, Mass.: M.I.T. Press, 1971.

Neumeyer, F., "The Employed Inventor as Subject of Legislation - an Ideological Survey," Industrial Property, 1971.

Schade, Hans, "Employees' Inventions - Law and Practice in the Federal Republic of Germany," Industrial Property, Sept. 1972.

Scherer, F. M., Industrial Market Structure and Economic Performance, Chicago: Rand McNally, 1970.

Simon, Herbert A., "Theories of Decision-Making in Economics and Behavioral Science," American Economic Review, 1959.

Stedman, J., "Employer-Employee Relations," in Neumeyer, F., The Employed Inventor in the United States, Cambridge, Mass.: M.I.T. Press, 1971.

CHAPTER 3

Aspects of the U.S. and West German Patent Policies

The patent system was devised to promote invention, to encourage diffusion through disclosure, and to encourage the investment necessary for commercial utilization of the invention. Concomitantly, the system has been attacked as granting monopolies and impeding technological growth. The dilemma remains, and one must balance the benefits of innovation stimuli against their costs.[1]

Let us realize that, in general, innovation results from searches for better solutions.[2] On the firm level, the stress leading to this search behavior is usually competition. The patent system should serve as an incentive so that the search leading to innovation will be protected and rewarded.

The purpose of this chapter is to present some of the principal aspects of West German patent activity and patent law, and, in order to provide a baseline, the German patent law will be compared with comparable aspects of the U.S. patent law. In no way do I mean to imply that either law is superior.

Rather, I believe we can increase our perspectives and our choice of alternatives if, indeed, we examine portions of these comparable laws. It is with this in mind that I present Chapter 3.

Growth of West German Patents in U.S.

The U.S. patent office review gives an analysis of the annual foreign share of selected countries as shown in Exhibit 3-1. Note that West Germany, of all the foreign countries has had the largest share of

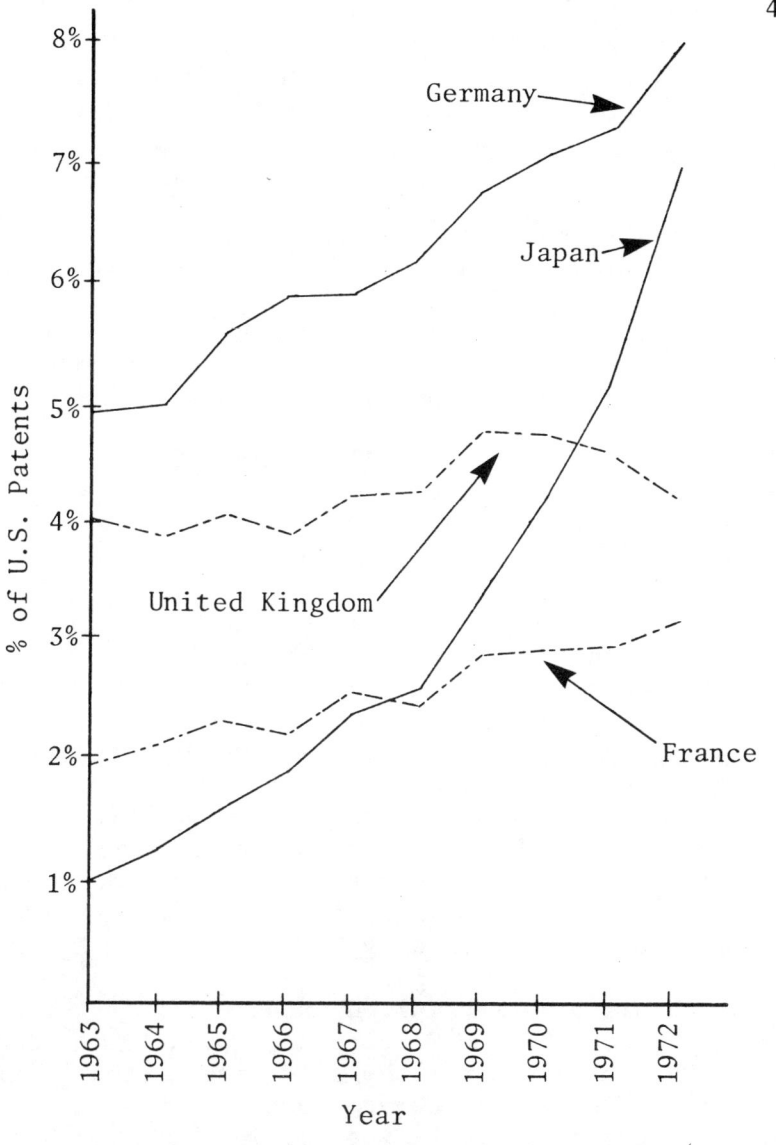

Annual Foreign Share of U.S. Patents
From Selected Countries*

Exhibit 3-1

*Technology Assessment and Forecast, Early Warning Report, U.S. Patent Office, December, 1973 (1973 Figures are Estimated at Approximately the Same as Those Shown for 1972).

U.S. patents, and that the percentage continues to increase. The United States patent activity in foreign countries is also on the increase and we see an interesting proliferation of technology.

We shall now dig a bit deeper to try to understand what lies behind those figures. Exhibit 3-2 shows the distribution of U.S. patents issued to residents of West Germany. The figure shows 2,891 patents issued from 1970 to 1973 in the category of "chemistry, carbon compounds." The second largest category contains 674 patents and is in the category of "drugs, bio-affecting and body treating compositions." The third and fourth categories are also chemically related, namely, coating processes, and compositions, catalysts, non-resinous mixtures. Therefore, if one combines the first four categories, one sees that 4,745 patents have been issued to West German residents in this reporting period in chemically related fields.

It is true, of course, that numbers of patents as a measure of technological activity has certain limitations. Clearly, we are not taking into account the varying importance of each patent. However, patent activity is one measure that gives an indication of national emphasis pinpointing high activity areas in the spectrum of technology.

Exhibit 3-3 is compiled identically to the previous one but is for U.S. patents issued to residents of the United States for the same time period. This exhibit enables the reader to compare areas of technological activity. By examining both exhibits, it is seen that chemistry is also highest for the category of U.S. patents. However, beyond this point, some interesting divergencies occur.

The second highest frequency of patent activity for U.S. is in the field of communication--electronics. This field is only the ninth highest for Germany.

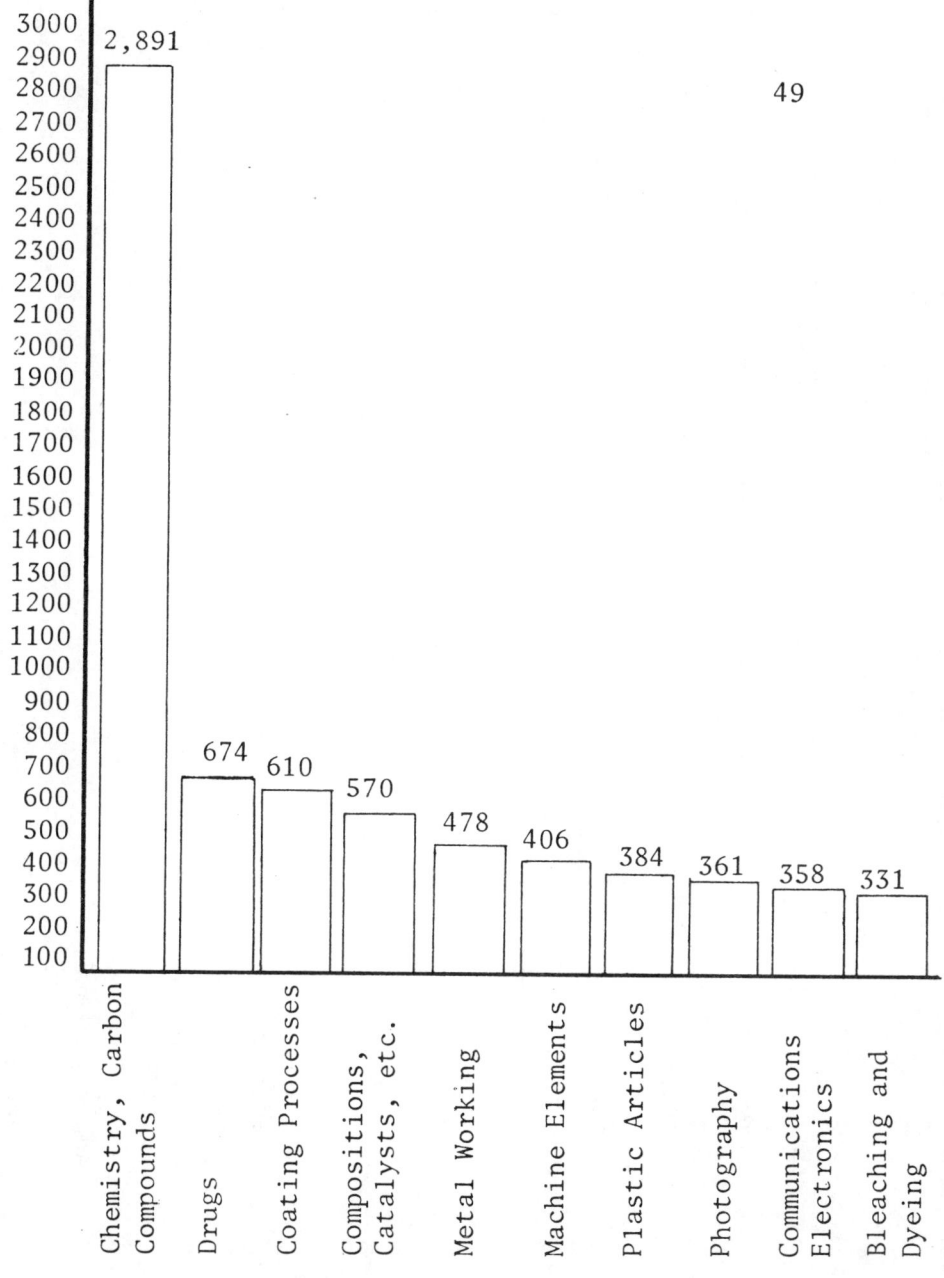

U.S. Patents Issued to Residents of West Germany 6/30/70 - 6/30/73*

Exhibit 3-2

*From, Technology Assessment and Forecast, U.S. Dept. of Commerce, Patent Office, Third Report, June 1974.

50

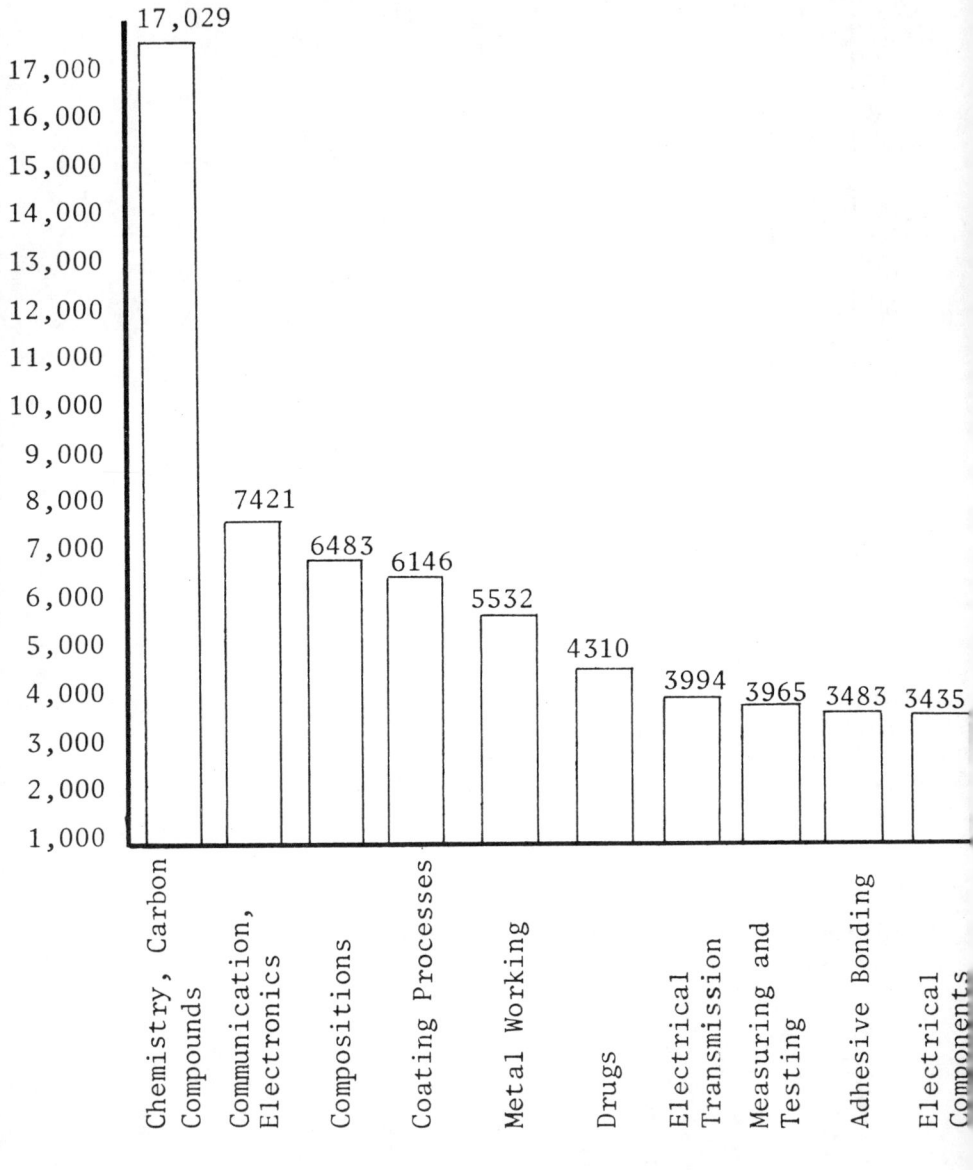

U.S. Patents Issued to Residents

of U.S. 6/30/70 - 6/30/73*

Exhibit 3-3

*From, Technology Assessment and Forecast, U.S. Dept. of Commerce, Patent Office, Third Report, June 1974.

Some other striking differences between West Germany and U.S. technological activity can be seen in other electrically related fields. In the top ten categories for the U.S., we see not only electric communication, but we also see electrical transmission, and electrical components (relays, capacitors, etc.). These last two categories do not make West Germany's top 10 list, but they are not far from it.

The number of West German patents in photography, (Exhibit 3-2) ranks in the eighth highest category, while this item does not make the U.S. top ten.

The next exhibit (3-4) examines the top 10 U.S. patentee companies for 1975. General Electric heads the list with 839 patents granted for the year 1975.* This finding is consistent with the previous exhibit which shows the U.S. emphasis on patents in the fields of electrical communication, electrical transmission, and electrical components. Note that G.E., A.T.& T., Westinghouse, I.B.M., and R.C.A. all appear in the top 10 patentee companies list adding further evidence of U.S. patent dominance in electrical devices.

Other companies such as General Motors, Dupont, Dow, and Phillips are also strong in frequency of patent activity. It is interesting that Kodak was the eleventh ranked patentee company. No other photographic company appears until Polaroid which ranks 37th in U.S. patentee companies for 1975.

---

*The original basis for the General Electric Company was Thomas Edison's patent called, "An Electric Lamp for Giving Light by Incandescence." (The highest number of patents issued to an individual in the U.S. went to Thomas Edison and the total is 1,093.)

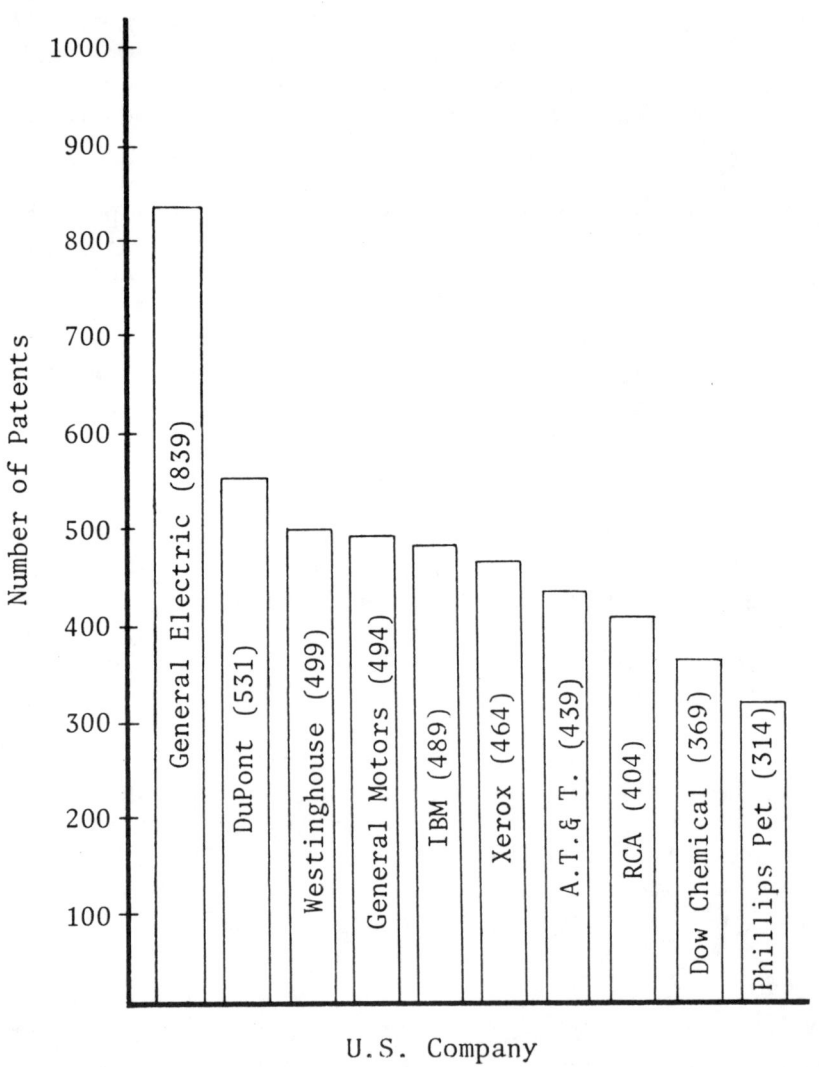

Top Ten U.S. Patentee Companies (1975)

Exhibit 3-4

A recent study in West Germany found that many firms are somewhat reluctant to file process patents because they prefer to keep the new process a secret within the firm. As in other countries, once a patent has been granted, it becomes public knowledge.

West German Broad Coverage vs. U.S. Specified

In West Germany, less than a third of patent applications are approved, compared to 70% in the United States. As might be expected, German patents are far likelier to stand up in court.

There are three reasons why, in West Germany, fewer than a third of the patent applications mature into patents, compared to 70 percent in the United States:

1. Searches and examinations in the U.S. are not made as thoroughly as in Germany.

2. The German Law provides for opposition proceedings permitting the public, in particular competitors, to take part in the official examination procedures. It is quite obvious that the competitors seek to destroy applications conflicting with their interests.

3. The standards applied to inventive merit are higher in West Germany than in the U.S.

In Germany, applications are judged by the following criteria:

(a) novelty;

(b) progress in the art;

(c) inventive merit.

Many applications are rejected on the grounds of lack of inventive merit, as compared to the other two categories.

Though the U.S. Law has the same criteria, the requirements for inventive merit are not as high as in Germany. There are even lawyers who maintain that the inventive merit plays only a subordinate role in the U.S.

In other words, a patent is said to be granted in the U.S. if proof can be furnished to the U.S. Examiner that the invention is new, constitutes a technical improvement and has, to a certain extent, inventive merit. The high standards applied to inventive merit in West Germany are often the subject of discussions and sometimes critically commented on.

One reason for the high U.S. patent nullification rate is that U.S. patents are very specific in comparison to West German patents which are based on more general principles. In the U.S., clear-cut wording with all components is used to describe the invention. The patent is so specific that it is relatively easy to nullify.* A change in any of the portions can be grounds for a patent suit.

The West German system uses general terms which are broader. This is in favor of the patentee since it becomes more difficult to nullify. Consider the two examples in Exhibit 3-5 A&B.

---

*The fact that the U.S. patent claims are more specific means that it is easier in the U.S. to determine an infringement, although the patent becomes easier to nullify.

Exhibit 3-5A

Comparison of U.S. and West German Patent Claim System
_____

Example A. Caliper and Friction Pad Support Means
_____

	U.S. Patent Claims		West Germany Patent Claims
Line 1	1. In a disc-brake system, the combination of a brake disk___ ---	Line 1	1. Spring hold and device for the brake jaws of a partially___ ---
			2. ---
	2. ---		3. ___ on the brake saddle from the bridge
	3. ---	Line 31	of each spring.
	4. ---		
	5. ---		
	6. ---		
	7. ---		
Line 138	8. ___ against said disc independently of fluid activation of said cylinder.		

Exhibit 3-5B

Comparison of U.S. and West German Patent Claim System

Example B. Dual Compartment Master Cylinder Liquid-
            Level Indicator

	U.S. Patent Claims		West Germany Patent Claims
Line 1	1. The combination, with receptacle means having a pair of liquid--- ---	Line 1	1. Device to indicate the level of liquid in containers with several---
	2. ---		2. ---mounted on the screw-top when the pressure plate is lifted by the float.
	3. ---		
	4. ---	Line 26	
	5. ---		
	6. ---		
	7. ---		
	8. ---		
	9. ---		
Line 58	10. ---provided with a lamp energized by said circuit means.		

Example A shows 8 claims and 138 lines for the U.S. patent. It goes into great detail and becomes relatively easy to overturn. The identical West German patent has only 3 claims and 31 lines. It gives broad coverage and would be difficult to nullify.

Example B in Exhibit 3-5 shows 10 claims and 58 lines for the U.S. patent while the West German patent for the identical part has only 2 claims and 26 lines.

A Fortune article pointed out the following in regard to U.S. Patents.

> The most seasoned patent lawyer is frequently unable to predict with any confidence whether a multi-million dollar investment in a new product will be protected by the patent law.*

Patent Life and Petty Patents

The patent life under the West German system is specified as follows:

> The term of a patent shall be eighteen years, beginning with the day following the filing of the application.**

This can be compared with the U.S. patent life of 17 years, although the U.S. patent life starts from

---

*The article describes the many pitfalls of the current U.S. patent system. Robert G. Hummerstone, "How the Patent System Mousetraps Inventors," Fortune, May 1973.

**German Patent Law, Revised, 1968, Part 1, Section 10.

the granting of the patent. In essence, the U.S. patent life is more like 20 years from the filing date since it usually takes about three years to receive the patent.

The petty patent system of West Germany grants from a three to a six year patent on minor inventions and improvements. This has an interesting appeal since it serves to prevent the extension of monopoly power for decades.

Petty patents in West Germany are registered at the Patent Office without examination procedure. Their maximum life is 6 years. To the holder, the West German petty patents afford the same protection as regular patents.

In the event of an infringement of a petty patent, the infringement court examines the petty patent as to patentability. The requirements as to petty-patent patentability are substantially lower than in the case of patents. In contrast with the West German Patent Law, petty patents can be granted on products only and not on processes. One must realize that the formation of a petty patent system raises the issue of determining which are major and which are minor patents.

First Applicant Principle

Only the United States and Canada have a system of priorities in determining who should receive the patent. In West Germany, the right to the patent shall belong to the inventor provided he was the first to apply for the patent. Consider the actual wording of the West German patent law as follows:

> If several persons have made the invention independently one of another, the right to the patent shall belong to the

person who was the first to lodge an application in respect of the invention at the Patent Office.*

This concept is quite different from the patent law in the United States. The U.S. law states that if someone can prove that he had the concept first, even though he filed later, the first concept inventor is to be awarded the patent.

The U.S. law is obviously more difficult to enforce as examiners must determine who has the first conception date. There is a certain inherent fairness in the U.S. law although it is more difficult to administer. An inventor must keep careful, well documented records, to prove the inception of the concept. In addition, the would-be inventor may not receive a patent if another person can prove that he started work on the concept prior to the conception date of the applicant. The German law assumes that the first to invent will also be the first to file for an application.

Patent Importance & Diffusion Related to Firm Size

A recent important study in Germany examined 1239 patents.** As a result of its investigations the study states the following:

---

*German Patent Law, Revised 1968, Part 1, Section 3.

**Info-Institut fur Wirtschaftsforschung, Munich, 3 April 1974, Patent & License Policy, Inst. for Economic Research.

> The smaller manufacturers in this study attributed no importance to the patent protection and the middle sized companies, only slight importance. . . .

Exhibit 3-6 shows the relationship between firm size and the influence of patent protection. The companies with an annual volume of $4 million to $20 million stated that for the 159 patents examined, the companies would have embarked upon 97% of these discoveries without the benefit of patent protection. These companies claimed that only 3% of the discoveries would not have been pursued without the benefit of patent protection.

According to the responses of the large corporations, they would not have made half of their discoveries without patent protection. To quote again from the IFO report:

> It was stressed again and again by the large manufacturers that the input of means into R & D can only be justified if a corresponding profitableness can be expected in the future. To be sure, nobody can diminish the risk of failure from the R & D efforts. This risk is diminished through the protection of the patent law.

The same IFO study showed that the period of time between the discovery and the first economic use, will vary with the size of the firm.* This is shown in Exhibit 3-7.

Exhibit 3-7 shows that in the smaller companies, 40% of the patents examined had economic use in the

---

*IFO Study, (Institut fur Wirtschaftsforschung), München, 3 April, 1974; Patent and License Policy.

Company Size, Annual Volume	N (Patents Examined)	The percent of actual discoveries co. would have embarked upon without patent protection	The percent of actual discoveries co. would not have embarked upon without patent protection
$ 4,000,000-$ 20,000,000	(159)	97%	3%
$ 20,000,000-$100,000,000	(273)	92%	8%
$100,000,000-$250,000,000	(229)	90%	10%
> $250,000,000	(539)	47%	53%

Influence of Patent Protection

According to Company Size*

Exhibit 3-6

*IFO, Munchin, 3 April '74, From Paper entitled, "Patent and License Policy", Institute fur Wirtschaftsforschung, Munich.

Company Size	N	Period of time between discovery and 1st economic use (in years)			
Annual volume in U.S. $ (millions)	(Patents examined)	Up to 1	Up to 2	Up to 3	Up to 4
4 - 20	(159)	40%	70%	78%	87%
20 - 100	(273)	32%	58%	66%	70%
100 - 400	(229)	26%	55%	65%	69%
400 & over	(539)	15%	33%	43%	50%

Exhibit 3-7

Period of Time Between Invention and Use

According to Company Size

1st year, only 15% for large companies. A partial explanation for the difference is that large corporations are concerned with more complex technology, and we can therefore expect a longer lag time between discovery and economic use.

Continuing to examine Exhibit 3-7, note that at four years, the smallest companies have made economic use of 87% of their patents, while the large companies have still only received economic use of 50% of their patents. Another partial explanation of these findings is that larger companies can afford and do, indeed, show high patent activity while realizing economic use from only a portion of their patents.

If we combine the findings shown in Exhibit 3-6 with Exhibit 3-7, we see that while the smaller companies in West Germany say they would innovate without patent protection, they do, however, obtain patents and do quickly convert their discoveries to economic use. On the other hand, the larger companies say they would be less likely to invest R&D funds without patent protection. But when they do invest, they take relatively long periods of time from invention to economic use. In large companies, the long time spans involved substantial investments and the patent system helps to reduce the risk of that investment.

Based on these findings one could hypothesize that larger companies obtain patents on marginal inventions in order to discourage competitors. The data seems to indicate that smaller companies cannot afford such a practice, and indeed convert 87% of their patents into economic use within four years. As might be expected, the medium-sized companies fall somewhere in between these two extremes. In general, other studies have shown that industries with high investments or value-added, account for more patents than those with low investment or value-added.

The lag between invention and innovation is

often significant. Some inventions constitute major breakthroughs while others are more routine. The ball point pen, the self-winding watch, and automobile power steering each took six years from invention to innovation. Inventions including freon refrigerants, the long-playing record, and plexiglass each took three or less years from invention to innovation. The data in this chapter indicates that the larger companies take longer from the time of invention to the marketed product. This may in part be due to the fact that these large companies are working on more complex technologies and partly due to the inertia of large organizations.

Compulsory Licensing and Fees

A fundamental difference in West German and U.S. patent laws concerns general compulsory licensing. In the U.S., general compulsory licensing is not allowed, while in West Germany it is an accepted feature of the patent law. Compulsory licensing can be involved when a patent holder fails to utilize his invention within a specific period of time, as well as when the patent recipient is using the patent to restrict supply excessively. In West Germany, the compulsory license will be invoked only when it is in the public interest. In actuality this has been tried 25 times since 1950, and, in each case, the parties came to a private agreement.

Although the U.S. Congress has not seen fit to enact such laws, the compulsory licensing has been specified in more than 100 antitrust cases, helping diminish some of the social costs of the patent system.

The issue of compulsory licensing has caused more heated debate than perhaps any other portion of the patent law. There are special statutes in the U.S. that provide for compulsory licensing, such as

the Atomic Energy Act, Plant Protection Act, and the
Clean Air Act. Here the view seems to be developing
that compulsory licensing of some sort is inevitable.*
However, the incentives of the patent system must
still be maintained.

An annual registration fee in West Germany dis-
courages individuals or companies from obtaining
patents and keeping them without their use. In the
U.S., once a patent is obtained, there are no annual
fees, and the critics voice concern over the social
costs of acquiring patents on products that are then
"shelved" and not made available to the economy in
general.

The West German law has a free system so that the
penalty for non-payment is the premature termination
of the patent. The invention is, of course, then
available for use by any interested parties.

The actual wording of the West German patent
law is as follows:

> . . . at the commencement of the third
> and of each subsequent year following the
> day after the applications has been filed,
> there shall be paid a renewal fee as pre-
> scribed in the table of fees . . .

---

*These remarks on compulsory licensing are from
a talk by Dr. John Stedman, Professor of Law, Univ.
of Wisconsin, formerly General Counsel to the Senate
Sub-committee on Patents, Trademarks & Copyrights.
The talk was entitled "Antitrust in Times of Economic,
Technological and Ecological Crises--the Patent-
Antitrust Crises." This talk was presented at a
symposium at the Pittsburgh Antitrust Inst., Nov. 1974
and published in the Univ. of Pittsburgh Law Review,
1975.

The renewal fees payable in respect of
the third and of each subsequent year shall
be paid within a period of two months of
their due dates. If the fees are not paid
within this period, the prescribed additional fee shall be paid in respect of the
delay in payment. At the end of this delay
the Patent Office will notify the applicant
or the patentee that the application is considered to be withdrawn or that the patent
will lapse . . .*

Mansfield points out that such a provision serves
to weed out worthless patents.** Scherer argues that
this provision of the law forces the patent holders
each year to reassess whether it is worthwhile maintaining their exclusive rights.*** <u>Under this scheme
fewer than 5 percent of all German patents remain in
force for the full term.</u>****

With the West German annual fee system, it is
found that half of all patents are dropped after 6 to
7 years by discontinuing payment of the annuities. As
I inquired among the companies, I found that the usual
practice is to check at regular intervals to determine

---

*Part 1, Section 11, West German Patent Law,
Revised '68.

**Mansfield, E. "Technological Change," Norton,
N.Y., 1971, p. 132.

***Scherer, F.M. "Industrial Market Structure and
Economic Performance," Rand McNally, 1970, p. 305.

****P.J. Federico, "Renewal Fees & Other Patent
Fees in Foreign Countries," Study No. 17 of the Subcommittee on Patents, Trademarks, & Copyrights, U.S.
Senate (Wash. 1958).

whether or not it is worth keeping the patents alive. Since a firm with many patents finds these costs significant they will let those patents expire which are recognized as being insignificant. Exhibit 3-8 shows the annual increasing fee that must be paid to maintain the patent. Although the individual charges may appear somewhat low, most products and processes are surrounded by many patents and the costs to maintain this exclusive right do tend to become significant.

This system, while eliminating worthless patents from the registry, has only slight effect on reducing the amount of monopoly power based upon important patents. The holders of significant inventions meet the necessary payments. This aspect of the law is of some value but should be considered as one positive element but not of undue importance when taken by itself. However, when considering the early portions of the chapter and the U.S. Patent Office's difficulties, imagine the significant savings that would result if this system meant that 50% of U.S. patents were dropped after 6 to 7 years.

There has been some discussion in the U.S. Congress concerning the establishment of maintenance fees of $500, $1,000, and $1,500 payable on the 7th, 10th, and 13th anniversaries of the term of a patent.

Let us see if we can put some of this patent information into perspective. There are many similarities between West German patent law and the United States statutes. In both situations the granting of a patent is to allow a monopoly. The patent monopoly is different from most other monopolies.3 In some respects the patent monopoly goes further than monopolies of the public utility type. For the patent monopoly is not limited to a reasonable return on investment, and, in fact, the inventor is offered extraordinary profits as an incentive.

Year Following Filing of Patent Application	Amount in $'s
3	20
4	20
5	32
6	50
7	70
8	100
9	130
10	160
11	210
12	270
13	330
14	400
15	470
16	540
17	610
18	680

Exhibit 3-8

West German Annual Patent Fee Table

In other respects, the monopoly is rather limited. The patentee must meet the tests of novelty and usefulness in order to obtain a patent. He is limited to specific time periods (18 years in West Germany, and 17 years in the U.S.).

The patent monopoly is limited not only to time periods but also to geography. For these monopoly privileges, a fee is paid--once in the United States and on a continuing basis in West Germany.

The effects a German patent and a U.S. patent have relative to third parties are substantially equal:

1. The patent holder may prohibit third parties to make the subject matter of the invention.

2. The patent holder may license a third party to use the patent. This is usually done in a license agreement.

3. Should a third party make unauthorized use of a patent, thereby committing an infringement, the patent holder may claim damages for past infringing acts and has the right to an injunction, which means that the third party is compelled to desist from its infringing acts in the future.

In this way, competition can be prohibited both in the U.S. and in Germany. This is the inventor's reward for his invention.

There are also regulations on the abuses of the patent monopoly. In West Germany, if the patentee is deemed to be too restrictive, compulsory licensing results. In the U.S., if abuses are excessive, the government takes appropriate antitrust action.

The Euro-patent

The Euro-patent has come about due to several factors:*

1) There is a continual increase in the territorial enlargement of markets. In nearly all cases, markets for new products and processes cover more than one national market in Europe. This increasing interpenetration of markets is forcing industry to seek patent protection in many countries. The large number of individual national patent applications and the widely differing examination procedures, have involved industry in considerable expenditure of time and money.

2) As investments for new products increase, the need for geographically broader patent coverage becomes essential. The Euro-patent will probably be substituted for the national one, since the grant of a Euro-patent offers the investor wider territorial protection. Those inventors and corporations with a national patent who are looking for an investor will soon discover that finding one solely on the basis of a national patent will turn out to be practically hopeless.

3) Due to the rapid development of technology, and the ever increasing number of patent applications, national patent offices are confronted with almost insurmountable technical and organizational difficulties.** Even though the patentability criteria

---

*Panel, Francois J., "Future Patent Policy of European Industries," International Review of Industrial Property and Copyright Law (IIC). Max-Planck Inst. Weinheim, Germany, Vol. 5, No. 2/1974.

**"The New European Patent Law," Commission of the European Communities, Directorate - General for Press and Information, Luxembourg, Sept. 1973.

of individual countries often coincide, a given invention which is the subject of a number of national patent applications has to be examined independently by each of the national patent offices concerned.

4) It is expected that the Euro-patent will cost about as much but no more than an extension to three countries today. The Euro-patent will be cost effective for those enterprises applying for patents in more than three countries.

The European Patent Convention (agreement) was signed on October 5, 1973 by West Germany and 13 other countries. The ratification procedure shall be concluded by 1976 at which time the Euro-patent shall go into effect.

Convention I establishing this European system for the grant of patents provides for a European patent organization with its headquarters in Munich and a branch office at The Hague. The Euro-patent will be granted centrally and will essentially consist of a "bundle" of individual patents with a term of 20 years from the application filing date. One may still be allowed to take out an individual country patent such as a German patent, or a French patent.*

The official languages of the Euro-patent office are English, French, and German. Exceptions to these language arrangements are provided for in the regulations. Applicants who have their headquarters or

---

*A second Convention (Convention II) is planned for the future under which the EEC will be regarded as one territory. The EEC patent to be created by this Convention will differ from the Euro-patent in that it can be granted only for the EEC as a whole, and will supercede individual country patents. This system differs from the "bundle" of patents under the Euro-patent.

place of residence in a contracting country in which
another language is used officially may file their
application in the language of that country. However,
within a period of three months, they must submit a
translation of the application in one of the three
official languages of the European Patent Office.

The probable effects of the Euro-patent system
are presently being examined.* It is expected that
many changes will accrue:

  .. Applicants who presently apply for
several patents (one patent in their
own country and three other national
patents) will utilize the Euro-patent,
thereby extending their coverage at
reduced costs and increased protection.

  .. Certain applicants who at present do not
apply for a patent at all, will be
attracted by the geographical breadth
of the Euro-patent.

---

*Panel, Francois J., "Future Patent Policy of
European Industries," International Review of
Industrial Property and Copyright Law, Max-Planck
Inst. Weinheim, German, Vol. 5, No. 2, 1974.
 Also see,
    Kolle, Gert, "The Patentable Invention in the
European Patent Convention," International Review
of Industrial Property and Copyright Law, Max-Planck
Inst. Weinheim, Germany, Vol. 5, No. 2, 1974.
Also see,
    Bruchhausen, Karl, "The Extent of Protection
of the European Patent," International Review of
Industrial Property and Copyright Law, Max-Planck
Inst. Weinheim, Germany, Vol. 5, No. 3, 1974.

.. As far as application for European patents from non-European enterprises are concerned, especially American and Japanese enterprises, there is no doubt that they will take advantage of the Euro-patent as early as possible. The American inventor prefers a unified European procedure instead of several parallel procedures, since neither he nor his American Counsel really know each countries' system. The American firm desiring patent protection in West Germany will be more inclined to apply for the Euro-patent, thereby receiving West German protection plus that of the remaining Euro-patent countries.

Chapter 3

Selected Readings

Cyert, Richard M., and James G. March, *A Behavioral Theory of the Firm*, New Jersey: Prentice-Hall, 1963.

Driffill, John, Carole Kitti, Mary Summerfield, and Charles L. Frozzo, "The Effects of Patent and Antitrust Laws, Regulations, and Practices on Innovation," Arlington, Virginia: *Institute for Defense Analyses*, 1976.

Hummerstone, Robert G., "How the Patent System Mousetraps Inventors," *Fortune*, May 1973.

Mansfield, D., *Technological Change*, N.Y.: Norton, 1971.

Scherer, F. M., *Industrial Market Structure and Economic Performance*, Chicago: Rand McNally, 1970.

Stedman, John C., "Patents and Antitrust - The Impact of Varying Legal Doctrines," *Utah Law Review*, No. 4, 1973.

Vernon, Raymond, *Manager in the International Economy*, New Jersey: Prentice-Hall, 1972.

CHAPTER 4

Funding for Technology: The Firm

The efficiency of a country's money and capital markets is instrumental in the allocation of savings to the most promising investment opportunities and in the growth and development of a viable economy. <u>The more varied the vehicles by which savings can flow from ultimate savers to ultimate users of funds, the more efficient the financial markets of an economy tend to be.</u>*

This chapter will focus on funding for technology in large, medium, small and new venture organizations in West Germany with comparisons to U.S. corporations.[1] It is recognized that innovation emanates from many size organizations, although differently from them. We shall first consider the large West German organizations, then the smaller companies, and finally the venture capital systems.

Centralization of Large Organizations[2]

In a recent analysis on corporate size, it was pointed out that 5 out of the largest 15 European Corporations are West German firms.** Another example

---

*See James C. Van Horne, "Financial Management and Policy," Second Edition Prentice-Hall, Englewood Cliffs, N.J. 1971, p. 292.

**Fortune, Aug. 74. Three out of the largest five chemical companies in the world are the West German firms of Hoechst, BASF, and Bayer.

75

of the considerable size of West German firms is seen in the chemicals industry. The firms of Hoechst and BASF exhibit high industrial concentration and both have surpassed Dupont as the leading world producer as measured by sales.

The number of companies listed on the West German stock exchanges have been <u>decreasing</u> continually since 1962. Exhibit 4-1 shows the number of companies for each of those 12 years. Although there were 643 companies in 1962, there are now 489, a decrease of 154 companies. At first this decrease seems to be a strange phenomenon considering the outstanding economic growth during the same period. By way of comparison, the New York Stock Exchange showed an increase of 383 companies for the same time period and the American Stock Exchange an increase of 439 companies.

When I asked for explanations for the decreased number of West German listed firms, the responses were generally as follows:

a) <u>Merger activity</u> - increased technology requiring increased combinations of skills and increased capital;

b) <u>Failures</u> - a certain number of companies are unable to survive the high technology, high capital investments required.

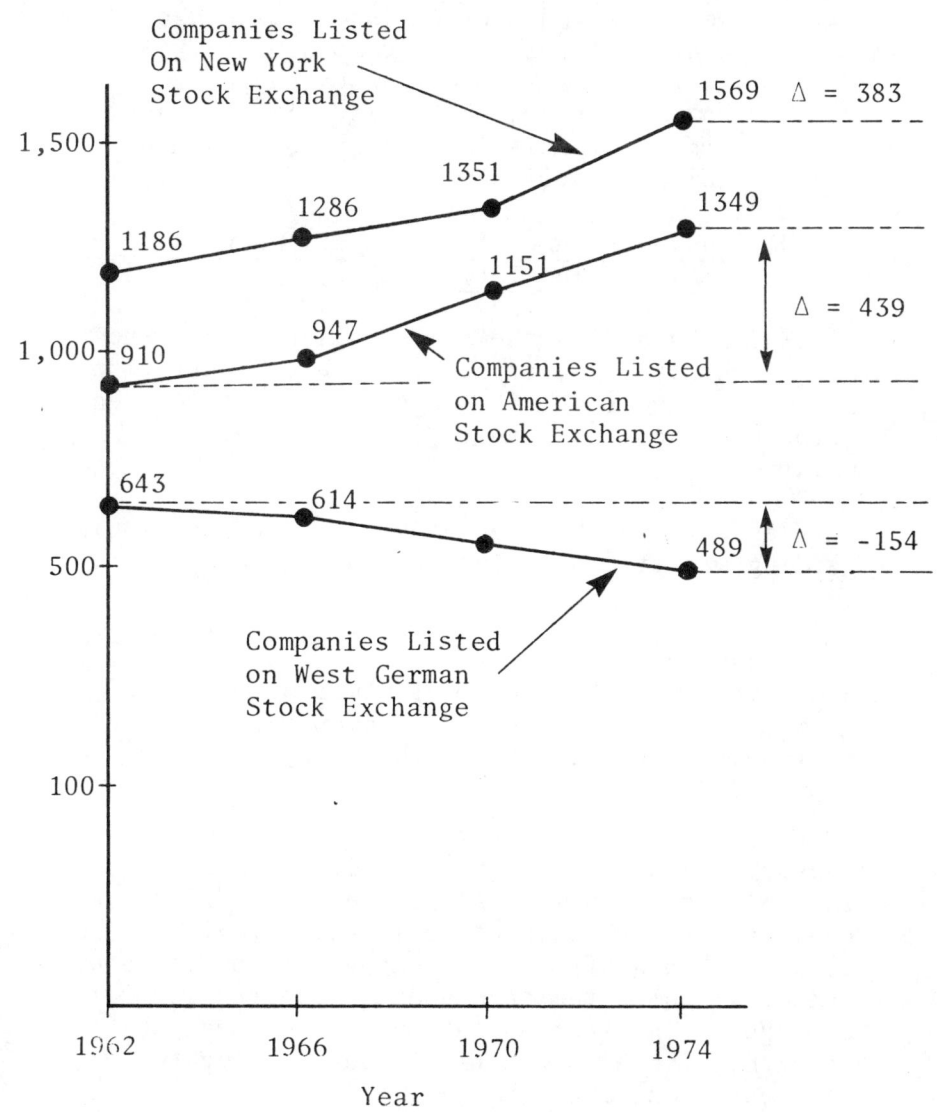

Companies Listed on West German
and U.S. Stock Exchanges
Exhibit 4-1

    c) <u>Low level of new entry</u> - Where the New York and American stock exchanges continually show new companies being listed, the West German exchanges do not. Instead, what exists is increased growth of existing companies with minimal new entries.

    d) <u>The existence of organizations with limited liability</u> (GMBH's).

    Notwithstanding the above explanations, the ultimate result is nevertheless high concentration of resources. Many economists argue that high concentration of resources results in high levels of innovation.*

    Jesse W. Markham recently stated the following regarding market concentration and innovation:**

---

    *Schumpeter argued that large scale establishments have come to be the most powerful engine of economic progress. Joseph A. Schumpeter, "Capitalism, Socialism, and Democracy." (Third ed; New York: Harper, 1950), p. 106.
Also:
    Galbraith supports the requirement for large business as a necessary prerequisite for todays technology as follows: "Associated with growth, as a goal of the technostructure is technological virtuosity," and "From these changes (sophisticated technology) comes the need and the opportunity for the large business organization." John Kenneth Galbraith, "The New Industrial State," Houghton-Mifflin, Boston, 1967 (p. 174).

    **Jesse W. Markham, "Concentration: A Stimulus or Retardant to Innovation," Harvard Reprint Series, 1975.

Some hold that concentration and large firm size are necessary to innovation, and that innovation is the most vital of competitive processes. An alternative view is that large firm size and concentration retard technological change. An evaluation of these two opposing views has great significance because a small difference in the rate of innovation may produce substantial differences in the rate of growth of the gross national product.

It is recognized that most of the major innovations do come from large organizations. Vernon points out that Dupont spent $2 million in 1935 on the development of nylon, and IBM in the billions for the third-generation computer.* These innovations could obviously not have emanated from small, or even medium sized companies. In another study, it was pointed out that there was clear evidence that their easier access to capital favors large companies' innovative activity.** It is also clear that not only does the scale allow such expenditures, but equally important, it allows time. Major innovations occur over long periods of time. Such innovations are not feasible for the medium or small sized company to undertake.3

In the steel industry, Adams & Dirlam reported that small fabricators were responsible for major innovations.*** The large steel companies, however,

---

*Vernon, Raymond, "Sovereignty at Bay," Basic Books, N.Y., 1971, p. 92.

**Ray, G.F., The diffusion of new technology, Natl. Inst. Eco. Review, May 69.

***Walter Adams & J.B. Dirlam, "Steel Imports & Vertical Oligopoly Power," American Economic Review, Sept. 1964.

refuted Dirlam's contentions. In actuality, there are undoubtedly contributions to innovation in all sized firms, but the contribution is usually proportional and obviously effected by the resources available.

The large corporations in West Germany generally fund their own technology. They reinvest portions of their earnings each year for the development of new products and processes. The chemical industry in West Germany gets no R & D money from the government. They do not need it, and they do not want it.

The companies in West Germany have been allowed to grow because there has been much less emphasis on antitrust than in the United States. Antitrust is at present regulated in West Germany by the "High Cartel Authority" (H.C.A.). West German antitrust law has been gradually tightened during the post war years. Before 1965, mergers had to be reported only if the merged firms controlled 20% of the market. The 1965 regulations require reporting of all mergers or purchases which result in payroll involving over 10,000 employees, gross sales of more than $125 million, or total assets of more than $250 million. The H.C.A. holds public hearings prior to such a proposed merger and supervises market practices of firms which dominate their respective markets. Similarly, it may prohibit acts and nullify contracts which it deems unfair or harmful. However, until about one year ago, there was little préssure against mergers. Furthermore, restrictions are generally directed against horizontal mergers, so we continue to see the rise of vertical mergers.

Another reason that the companies get larger is that in West Germany there are fewer spin-off companies than in the United States. The phenomenon of spin-off is essentially non-existent in Germany. It is hard to exactly determine why but several possible explanations seem plausible:

a) The culture is such that it would be considered a disloyal and ungrateful act for two or three engineers to leave one company and start a new company. This process is, of course, common in the United States and has often resulted in significant innovations. There have been some studies which show how the spin-off companies often far exceed the parent company in sales after a short period of time. The spin-off ventures in the United States are characterized by high growth and survival rates and by a high degree of technology transfer. Roberts and Cooper have studied such transfers and spin-offs in the Boston and Pal Alto areas, respectively.*

b) The original German company allows innovations to spring from within the company and through the use of the inventors law (see Chapter 2), the individual innovators are not forced to spin off, into separate organizations. There have been numerous cases in the United States where the spin-off resulted from organizational frustration, and it was only through the formation of a new company, that the innovator could hope to accomplish his goals.

c) The availability of venture capital in West Germany has been almost non-existent, so the financing necessary for the spin off company was just not available. This particular feature is changing with the formation of risk capital groups. However, it is doubtful as to whether these groups will encourage spin off organizations given the other factors dissuading them.

d) In West Germany there is a great deal less cultural flexibility. The individual does not readily move from a university to business, from a business to

---

*E.B. Roberts & H.A. Wainer, in "Proceedings of the 20th Natl. Conference on the Admin. of Research" (Denver Research Inst., Denver, Colorado, 1966), pp.81-92. A.C. Cooper, IEEE Trans Eng Mgt. EM-18, 2 (Feb. 71).

a university, or from public service to a university, etc. In West Germany, you choose your place, and you stay there. On occasion, an individual might switch careers once but with nowhere near the frequency as people in the United States do. This high structural viscosity acts as a deterrent to West German spin offs.*

One West German professor with whom I spoke and who has given the industrial scene extensive study, summed it up well as follows: "If one observes the U.S. industrial scene between 2 ten-year time spans, one sees enormous change. If one observes the West German industrial scene between these same two ten-year time spans, one sees less change, but merely the large corporations becoming even larger."**

Innovation from Medium and Small Sized Companies[4]

Since innovation springs from many sources, let us now consider medium and small size companies. Much of their funding comes from within as with large organizations. However, in order to provide an incentive for the introduction of new products and new processes from small and medium businesses, the West German reconstruction bank issues long term, low interest loans for just these sized companies. It is recognized that some innovation must come from other sources than the government and large businesses. It is also recognized that medium and small sized

---

*The concept of high structural viscosity in West Germany comes from Dr. Helmar Krupp, Director, Inst. for Systems Engineering and Innovation Research (ISI), Karlsruhe, Germany.

**Prof. Otto Poensgen, Univ. of Saarlandes, Saarbruken, Germany.

companies have difficulty in financing innovation. As
a result, there is an institution devised for precisely
that purpose.* (It performs other functions as well,
although this is its major activity.)

In order for a company to qualify for a loan, a
specific new product or process must be proposed. On
this basis, the companies have been receiving increased loans to finance these innovations. Exhibit
4-2 shows the domestic investment loans in 1962 at
230 million dollars, increasing to a figure in 1974 of
1.35 billion dollars. The dip in 1973 is "on account
of considerations of stability policy."** However,
the 1974 figures are beyond $1.3 billion. These funds
obviously provide no guarantee of future innovation,
however. With other portions of the innovative system
present, the addition of capital certainly increases
the probability of innovation.

Official funds from the Federal Budget as well
as from the Reconstruction Loan Corporation's own
funds are employed for financing loans. It is in this
way that it can offer long term low rate loans. (This
lending institution actually uses other banks as
intermediaries but that is mainly to aid in the security checks for the applicants).

The loans to medium and small businesses as
stated by the Reconstruction Loan Corporation are for
"investments for new products and services, fundamental rationalization, research and development. . ."
The guidelines were recently extended so that it is
now also possible to assist in the financing of

---

*Kreditanstalt, (Reconstruction Loan Corp.),
25th Annual Report, 1973, Frankfurt, Germany.

**Kreditanstalt fur Wiederaufbau (Reconstruction
Loan Corp.), 25th Annual Report, Frankfurt/Main,
Germany, p. 31.

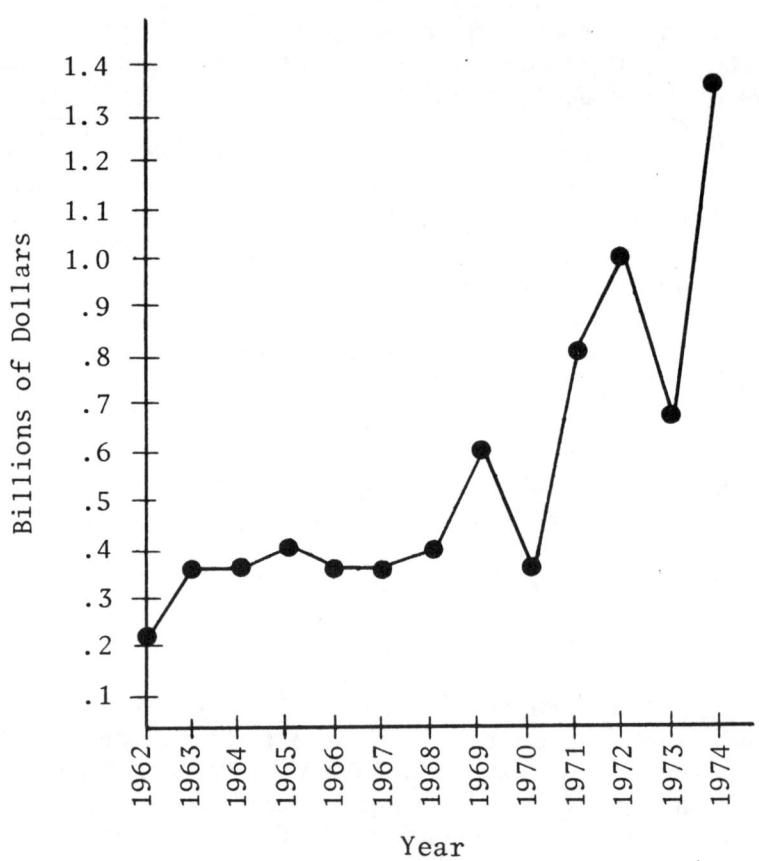

West German Domestic Investment Loans

Exhibit 4-2

investments aiming at a more efficient use of energy sources. The Reconstruction Loan Corporation states in its 1972 statement:

> Moreover, financing funds are provided which are not regionally committed, as for the introduction of technical innovations for utilizing modern machinery and equipment, for example, or for including new items into the product line. Loans were further granted for adopting research and development results . . .*

The granting of loans by category is as follows:

Medium Sized Companies		
Introduction of New Products	25%	
Introduction of New Processes	39%	67% for
R & D	3%	innovation
Location	27%	
Capacity Expansion	6%	
	100%	

It is interesting to note the emphasis on new processes (39% vs. 25% for new products). The medium sized companies are seldom granted a loan toward general capacity expansion (6%). The feeling is that clear purposes should be described. Note also that only 3% is loaned for R & D (actually 2% for research and 1% for development), the reasoning being that the company should be doing this R & D investing themselves with loans available during the expensive introduction period. It is also very difficult to finance the early R & D since so many projects are failures.

---

*Kreditanstalt fur Wiederaufbau (Reconstruction Loan Corp.), 24th Annual Report 1972, Frankfurt (Main), Germany.

The bank also wants to see the commitment of the company shown through its investment in R & D. During the introductory portion of the innovation, the credit is available with 39% going for new processes, and 25% for new products. The emphasis on processes is clearly a result of the high labor costs and tight labor market at that time.

In 1974, the Reconstruction Loan Corporation estimates that they will issue 8,000 loans for small and medium sized companies. Of these it is expected that the division will be 50-50 between small and medium businesses.

Risk Financing and Venture Capital

A source of innovation that has not been encouraged in West Germany until recently, are new ventures. On the overall figures this may not be a significant source for innovation, but it is nevertheless of some value, and, on occasion, could be of very high value. Until recently, there were essentially three basic ways to finance new technologies.

1) <u>Private financing</u>: There are a small number of innovations which are launched by private capital, either through the individual innovator, or through combinations of individuals interested in investing in new ventures.

2) <u>Corporate support</u>: The existing corporations, on occasion will support new technologies, so that an innovator may approach them for financial backing. In most situations of this kind, the parent company takes over control and the individual innovator does not retain independence.

3) <u>European Economic Development Company (EED)</u>:

The E.E.D. is a European venture capital company which supplies capital and managerial assistance to help outstanding individuals build new companies or develop existing ones, based on advanced products, processes or services, thus leading to capital appreciation for the owners of these businesses and for EED stockholders. Investment opportunities in any field of endeavor which is felt to be constructive and to possess substantial possibilities for growth are considered.*

The European Enterprises Development Company provides long-term capital and essentially takes the same risks as its partners in the venture. The amount of capital which EED invests in a single situation is flexible and is determined by the project's requirements. It not only supplies capital, but supplies managerial assistance as well. Although the EED does not seek control in the companies in which it invests, in certain cases, it will take over control for a limited time period. In Europe, the concept of venture capital is quite new, but, in Germany, pressure has been mounting for a separate German equivalent to the EED.

As an example of EED support in West Germany, the EED supplied capital for the establishment of a glass fiber reinforced tubing company. The firm produces glass fibre reinforced polyester tubes with special joints and bends mainly for the chemical industry. A second example of EED investment in West Germany, involved a company performing research, development, and manufacture of high power gears for transmission. The company has developed and is

---

*European Enterprises Development Co., Annual Report 1972, Paris.

producing a range of high power mechanical controllable gears and coaxial rotating ball gears for maximum transmission.

The EED actually supplies various forms of assistance in addition to the financing. It supplies assistance to companies for promoting new products. It has strong press coverage aiding the launching of marketing programs. As such, it performs an important role in the innovative system. It also effects innovation by introducing to their funded companies new products and techniques which help to complete their product range or to produce more economically. Thirdly, the EED serves as an information diffusion system and brings companies into contact with other companies often leading to technological transfers, and, in some case, to mergers.

However, the overall effect of EED on innovation in West Germany is minimal. The vast majority of the investments have been in other countries. The pressure has been increasing in West Germany for a national venture capital system.

4) West German Venture Capital[5]

The current thinking in West Germany is that none of these three techniques appear sufficient to serve as incentives for new or small businesses, as the economy of the country needs the value of many small and new firms to add to the current industrial base. The consensus is that the country needs a strong record of exports to be successful. This record can only be accomplished through continual growth of existing firms as well as the infusion of new firms.

The plan approved in early 1975 is to form an organizational model as shown in Exhibit 4-3. Each bank would contribute a certain amount of money so that the individual banks' risks become minimized.

89

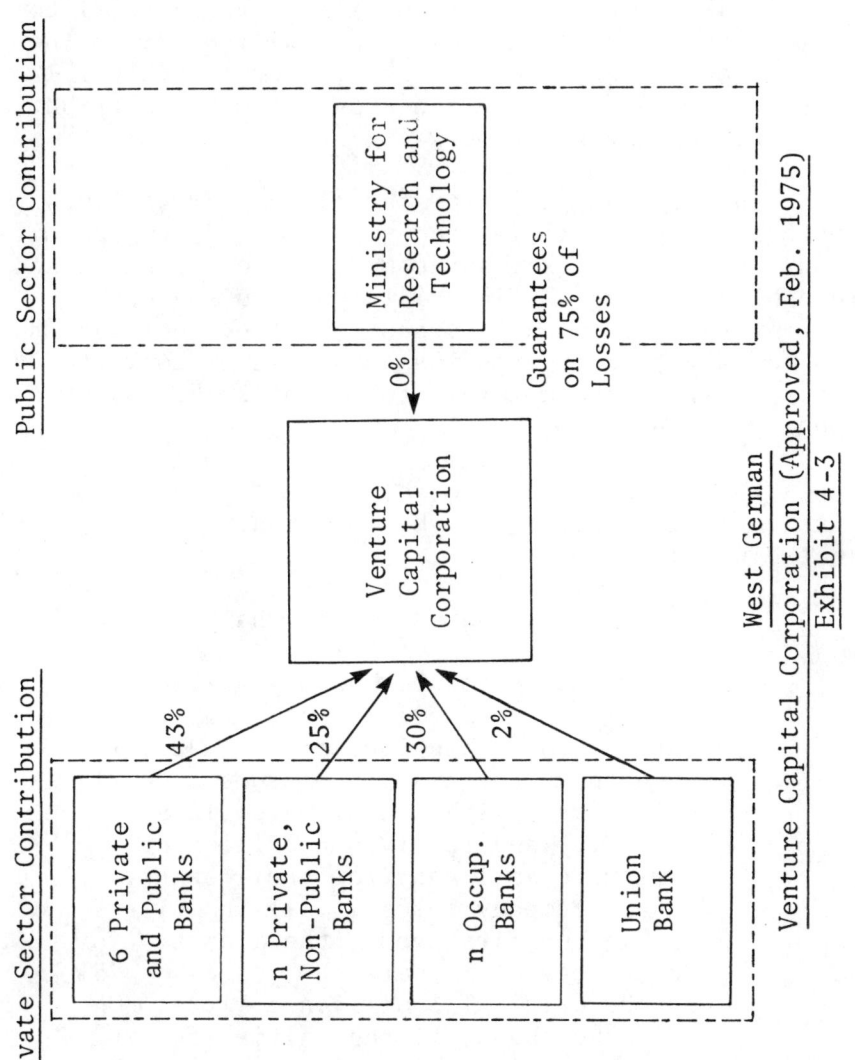

West German Venture Capital Corporation (Approved, Feb. 1975)
Exhibit 4-3

The plan is to start with a base of 6 banks contributing 43% of the necessary capital. As shown in Exhibit 4-3, the next portion of the capital will come from private banks and their share will be 25%. The Occupational Banks (such as The Farmers' Bank) will contribute 30%. Exhibit 4-3 shows a small contribution from the Union Bank, namely 2%.

The total capital for the first year will be approximately $4 million and will build up to $20 million. Exhibit 4-3 shows the Ministry of Research and Technology having no financial contribution but guaranteeing 75% of all losses. In the event that the new company's one year's losses are later converted into profit, the company is to pay back the invested funds plus interest.

Banks are, of course, hesitant to embark upon risk capital ventures and this hesitancy is alleviated in two ways:

a) the risk is shared by many banks; and

b) the Ministry for Technology will cover losses up to 75% of invested capital but not more than 100% of the capital invested. For example, assume a company starts with 3 million dollars initially and receives 6 million dollars more as loans from their banks. If the company loses 4 million dollars within the first year, the Ministry for Technology will share the losses by 75% of the original 3 million dollars. On the other hand, if the initial funding was the full 9 million dollars and the company lost 4 million dollars in the first year, then the Ministry for Technology would share the full losses of 4 million dollars by 75%.

The two basic models that would qualify for a risk venture are shown in Exhibit 4-4.

Essentially there are no limits as to the field of activity in venture capital groups, although this is still somewhat of a debate. One group, largely consisting of government officials, feels that the venture capital organization should finance only socially desirable products and processes. Another faction consisting largely of bankers, feels that the overriding criteria should be potential profit and growth. However, at present there is no limitation on projects, although the government influence would obviously encourage those projects that would fall closer in line with its current goals and objectives. These would include products for decreasing pollution, improving transportation, public health, etc. Since there is Federal participation in the supervisory board, the venture capital corporation would not allow socially undesirable products to be funded, such as high pollutant paints, dangerous toys, etc. To be economically viable, the basic criteria will be potential profit and social desirability.

Ultimately, the goal is to sell the participation in each company to other interested companies and make a profit. The Ministry for Technology provides insurance against aggregate losses, so either the Venture Capital Company as a whole makes a profit, or it makes a loss as a whole. What one must measure is the potential value of stimulating innovations. For it only takes a few innovations, as history has shown, to make a remarkable change in the growth of a company or country.

The most difficult aspect of the entire concept is the age old problem of management. The current plan is to employ approximately 20 people in the venture capital organization of which 10 would be engineers and economists and 10 would be assistants and secretaries. The major stumbling block is to find

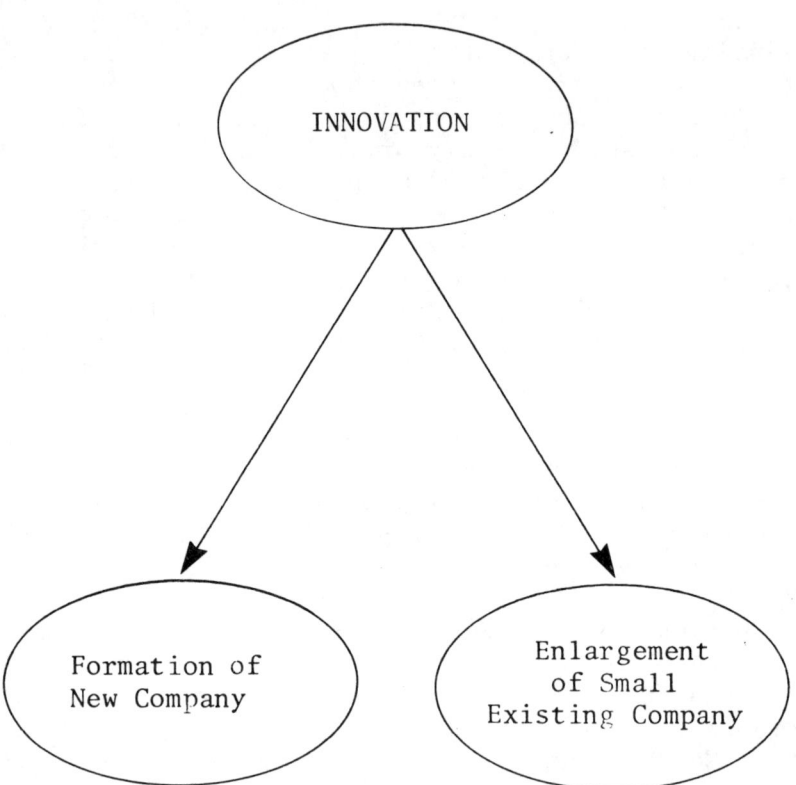

Qualification Model for Risk

Venture in West Germany

Exhibit 4-4

the combination of technical, economic, and managerial skills to manage such an organization. And West Germany, like many other nations, is faced with the severe shortage of well trained managerial personnel.

Assuming that the staffing is successful, the present plan is to evaluate 200 potential innovations in the first year, and to select 4 or 5. The financial volume for the average project will be $750,000 to $1,200,000.

The role of the government is not only to guarantee 75% of the invested capital against losses but also to supply a staff of free advisors even though the government itself has no direct investment.

In considering the funding for large, medium, small, and venture organizations it becomes clear that the functions of technology and finance do not have to be performed by the same organization. Scherer points this out well where he describes organizational separation.6 The importance of the financial community to innovation cannot be stressed enough.7 However, as important as this relation is, it is again one subsystem of the larger technological system. We have many examples of capital expenditures for innovation that have resulted in failure because other portions of the technological system were lacking. It is only through the combination of funding and through government, industrial, legal, and education efforts that innovation can take place.

Chapter 4

Selected Readings

Adams, Walter, and J. B. Dirlam, "Steel Imports and Vertical Oligopoly Power," American Economic Review, Sept. 1964.

Anthony, Robert N., John Dearden, and Richard F. Vancil, Management Control System, Homewood, Ill.: Irwin, 1972.

Cooper, A. C., "Spin-Offs and Technical Entrepreneurship," IEEE Transactions for Engineering Management, EM-18, 2, Feb. 1971.

Galbraith, J., The New Industrial State, Boston: Houghton-Mifflin, 1967.

Markham, Jesse W., "Concentration: A Stimulus or Retardant to Innovation," Cambridge, Mass.: Harvard Reprint Series, 1975.

Schumpeter, Joseph A., Capitalism, Socialism, and Democracy, New York: Harper, 1950.

Twiss, Brian, Managing Technological Innovation, London: Longman, 1974.

Van Horne, James C., Financial Management and Policy, Englewood Cliffs, New Jersey: Prentice-Hall, 1971.

Vernon, Raymond, Sovereignty at Bay, New York: Basic Books, 1971.

CHAPTER 5

Factors Related to Successful Innovations*

During interviews with research managers, project leaders, and engineers in 1974 I examined 32 innovations.[1] These are divided up into 11 successful, 11 unsuccessful, and 10 innovations in progress. This chapter will compare the successful with the unsuccessful innovations. The goal is to try to identify some factors that may contribute to success or failure and perhaps make some steps toward a better understanding of the innovation process in general.

These innovations are from West Germany and may serve the dual purpose of supplying the reader with insight into the West German innovation process and also information on the general innovation process.

Most innovations are relatively small steps forward as opposed to "major breakthroughs." It is incrementalism, that is, the accumulation of such technologies, that determines technological change. This study examines these types of innovations. It is expected that the results in this chapter are somewhat generalizable . Many portions of this book apply only to West Germany. However, some of this chapter's findings are found to be consistent with previous studies in other locations where innovation is taking place, hence, are believed to exist across cultures.

Project Descriptions

For purposes of this study, a successful innovation is defined as one in which a minimum of one man-

---

*Sections of this chapter have appeared as an article by me in I.E.E.E. Transactions for Engineering Management, Northwestern University, Aug. 1976

year of effort was expended between origin of the idea and the introduction to the market. This effort was to have taken place since 1970, with the marketed product showing all indications of continued success. The unsuccessful innovation was to have had a minimum of one man-year of effort exerted since 1970 and to have been stopped with no signs of continuation.

Exhibit 5-1 shows the 11 successful innovations from West Germany that I examined during 1974. These projects were selected by asking project managers to describe a recent, successful project as defined. The project may involve a process, such as item E in Exhibit 5-1, the super heated steam process. The project may involve a product, such as item A in Exhibit 5-1, the shock absorbent auto bumper. The exhibit shows examples of successful innovations from the automotive, electronics, and chemical industries.

Since the goal was to compare successful with unsuccessful projects, the next step of the research was to ask the appropriate project managers to describe a recent project failure. Many people felt that this information would be difficult to obtain since company spokesmen might be hesitant about speaking of failures. In actuality, that was not the case. The project manager often replied that he had many more failures than successes and did not hesitate in describing them.

Exhibit 5-2 shows the 11 unsuccessful innovations from West Germany that I examined during 1974. A previous study of mine cites the large number of non-technical failures.* Of the unsuccessful projects studied, more than 60% failed for non-technical

---

*A. Gerstenfeld, C. Turk, R. Farrow, and R. Spicer, "Marketing and R & D," Research Management, (New York: Wiley, 1969); A. Gerstenfeld, Effective Management of Research and Development, (Reading, Mass.: Addison Wesley, 1970).

A. Shock absorbent auto bumper

B. Electronic relay

C. Modified turbo engine

D. Dye for non shrinking wool

E. Super heated steam process

F. New front axle for auto

G. Improved T.V. circuitry

H. One piece side panel for car

I. New fuel pump membrane

J. Ignition discharge system

K. Thin film circuits

Exhibit 5-1

Successful Innovations Examined in

West Germany in 1974

A. Cam action spring for car door stop

B. Mechanical switch for tel. dialing

C. Friction gear box for automobile

D. Motorcycle straight cylinder motor

E. New process to print by chemical reaction

F. Better mechanism for auto mech. shifting

G. Non friction brake

H. Reactive dye for wool

I. Automatic ticket reader

J. Plastic milk bottle

K. Polarized headlights

Exhibit 5-2

Unsuccessful Innovations Examined in West Germany in 1974

reasons. There have been numerous other studies with similar findings.[2] A non-technical failure can be well illustrated by item K in Exhibit 5-2, the polarized headlights. With this device the driver can see the light from his own headlights clearly but by means of a filter he no longer has the glare from on-coming headlights. In order to polarize the light from the headlight, the device contains a series of thin plates arranged in such a way that the output light is polarized and becomes unidirectional.

This innovation was worked on for four years and the project team solved the technical problems only to have the product fail for other reasons. The reasons cited were that introduction of the product became extremely difficult. The legislation had to be changed. An additional problem was that polarized headlights on the first cars would be of no help since the other cars would not have a similar system. Therefore, until it became widely accepted it was of little value to the purchaser. Further development was recently stopped, and there is no expectation of continuation. This is another example in the long list of non-technical project failures which used scarce technical resources that could have been more wisely applied.

Demand Pull and Technology Push*

In today's environment one observes a cycle between engineering and marketing as shown in Exhibit 5-3. The cycle can go from engineering to marketing and back again to engineering, or it can go from

---

*The sources of innovation are carefully considered in a recent issue of Science; James M. Utterback, "Innovation in Industry and the Diffusion of Technology," Science, Vol. 183, 1974, p. 620-26.

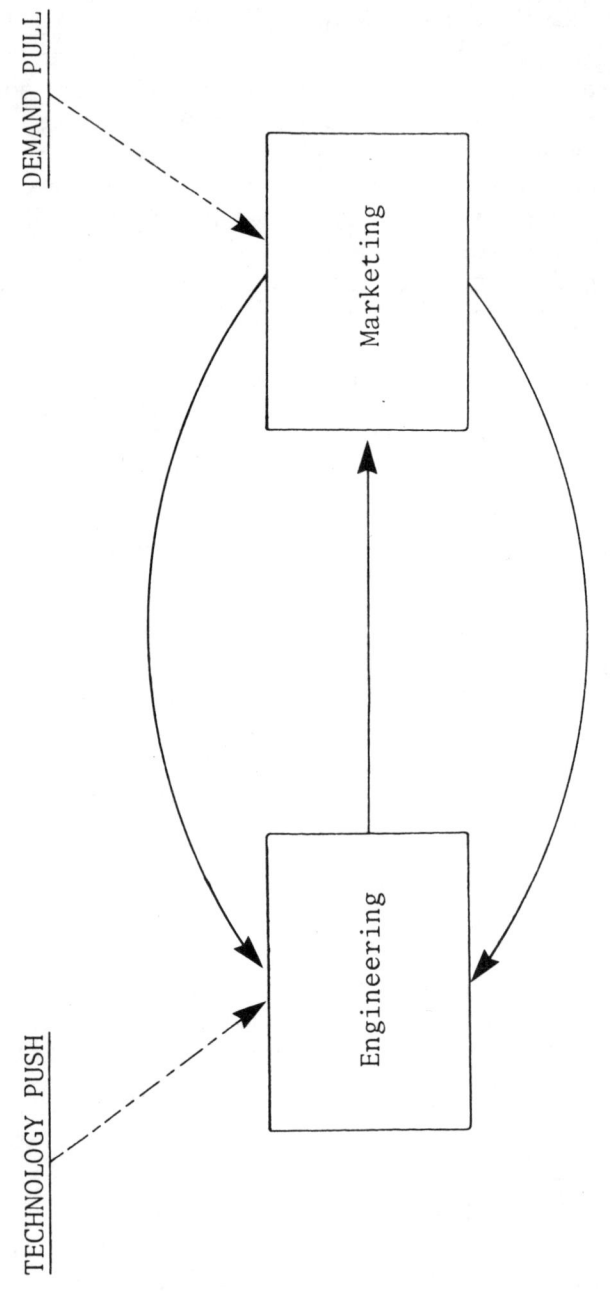

Continuous Collaborative Management

Exhibit 5-3

marketing to engineering and back to marketing. In
either case, the importance of a continuous collaborative relationship is paramount.3 In the event that
the origin is from engineering this is considered
technology push, as shown in Exhibit 5-3. On the
other hand, if the organization embarks upon a project
as a result of outside forces, this is labeled demand
pull as shown on the right side of Exhibit 5-3.4
If, indeed, the continuous collaborative relationship
is working, it should not make much difference where
the idea enters the system. However, it appears as if
firms have difficulties in maintaining such communications.

Within the past five years, there were three
separate independent studies published relating to the
subject of technology push/demand pull.5 These studies
all showed that more than two thirds of the successful
innovations were demand pull. One particularly interesting article stated the following:

> In our analysis of the whole set of 567
> innovations, we found that the vast majority
> of them--three quarters, in fact, were
> stimulated by a market demand or a production
> need. Only one fifth arose from someone saying, aha, maybe we can find a use for this
> technical idea.6

Therefore, the findings that I am about to
present should not be considered in isolation but
indeed as additional evidence of this concept. It
is the goal to try to get a somewhat better understanding of the innovation process and the unraveling
of the combination of factors that separate successful
from unsuccessful innovations. Exhibit 5-4 shows the
11 successful and 11 unsuccessful innovations that I
studied in West Germany in 1974. For the successful
innovations, 8 of the 11 were as a result of demand
pull. This means that the idea originated from beyond
the technical group. To be more specific, let us

Successful Innovations	Demand Pull / Tech. Push	Unsuccessful Innovations	Demand Pull / Tech. Push
Shock absorbent auto bumper	Demand Pull	Cam action spring for car door stop	Tech. Push
Electronic relay	Demand Pull	Mechanical switch for telephone dialing	Tech. Push
Turbo engine modification	Tech. Push	Friction gear gox for automobiles	Tech. Push
Dye for non shrinking wool	Demand Pull	Motorcycle straight cylinder motor	Demand Pull
Super heated steam process	Demand Pull	New process to print by chemical reaction	Tech. Push
New front axle for autos	Tech. Push	Better mechanism for auto mech. shift	Demand Pull
Improved T.V. circuitry	Demand Pull	Non friction brake	Tech. Push

Exhibit 5-4 (continued on next page)

Demand Pull vs. Technology Push for Twenty-Two West German Projects

Exhibit 5-4
(cont'd)

Successful Innovations	Demand Pull / Tech. Push	Unsuccessful Innovations	Demand Pull / Tech. Push
One piece side panel for car	Demand Pull	Reactive dye for wool	Tech. Push
New fuel pump membrane	Demand Pull	Automatic ticket reader	Tech. Push
Ignition discharge system	Tech. Push	Plastic milk bottle (inflatable)	Tech. Push
Thin film circuits	Demand Pull	Polarized headlights	Tech. Push

SUMMARY

Demand Pull = 8
Tech. Push = 3

SUMMARY

Demand Pull = 2
Tech. Push = 9

consider some of the innovations that were cases of demand pull. Exhibit 5-4 shows the shock absorbent auto bumper as demand pull. That is because the government insisted on this requirement and the technical group responded to this need. Similarly, the electronic relay was requested by a company and this request then became translated into an R & D effort.

On the other hand, technology push means that the origin of the idea comes from within the technical group. Exhibit 5-4 shows the turbo engine modification as technology push since the origin for the product was from engineering. It is also shown that technology push was the origin of the cam action spring, the non friction brake, etc.

Before returning to the analysis of these studied innovations, it is important to reiterate that the sample under study in West Germany was based upon small incremental innovations having an accumulative effect on national economy. Any conclusions we reach must be tempered by that fact, for writers have clearly stated that innovations causing major changes may well result from technology push.

Exhibit 5-5 shows the successful and unsuccessful projects in relation to this issue of demand pull and technology push. It becomes apparent as one studies Exhibit 5-5, that there is a complete reversal, and that success is related to demand pull and failure related to technology push. This does not mean that the firm should not embark upon technology push projects (for we cite several successes), however, the risks are higher.

Exhibit 5-5

Demand Pull-Technology Push*

Results

	Demand Pull	Technology Push	
Success	8	3	11
Failure	2	9	11
	10	12	22

The work done by Teubal in Israel showed that the proportion of failures in programs whose idea originated in R&D exceeds the proportion of failures in programs whose idea originates elsewhere.** (See Exhibit 5-6)

---

*The Fisher Exact Probability Test shows a level of significance = .025.

**Teubal, Morris; Arnon Naftali, and Manuel Trachtenberg, "Performance in Innovation in the Israeli Electronics Industry: A Case study of Biomedical Electronics Instrumentation Preliminary Findings," The Maurice Falk Inst. for Eco. Research in Israel, Sept. 1974.

Exhibit 5-6

Success and Failure of Programs,

by Origin of Idea (from Teubal)

Origin of Idea	Success	Failure	
R&D	0	15	15
Other	4	1	5
	4	16	20

Most firms do recognize the importance of innovations that result from demand pull.

Exhibit 5-7 compares the present study in West Germany with the 1974 MIT five country study (including West Germany), and the SAPPHO study in U.K. In each of these studies the successful projects were originally intended for a specific user or end product.

Exhibit 5-7

Characteristics Related to Project Success vs.
Failure in the Present Study Compared
with the Findings of Other Studies

	Present Study	MIT**	SAPPHO***
Projected intended for a specific user for end product	*	*	*

Level of Effort and Early Warning Systems

In general, the organization's choice in deciding how rapidly it will proceed in developing its product has been the concern of other writers.**** One graphic illustration of the concept is shown in Exhibit 5-8.

---

*For both the MIT and the SAPPHO findings there was a strong positive relationship between project success and specific user or end product ($p < .01$). For the present study in West Germany the same positive relationships appeared with a level of significance at $p = .025$.

**Utterback, J.M., T.J. Allen, J.H. Holloman, and M.A. Sirbu, Jr., "The Process of Innovation in Five Industries in Europe and Japan, IEEE Transactions on Engineering Management, Northwestern Univ., Feb. 1976.

***Robertson, A.B., B. Achilladelis, P. Jarvis, Success and Failure in Industrial Innovation: Report on Project Sappho, London: Centre for the Study of Industrial Innovation, 1972.

****F.M. Scherer, Industrial Market Structure and Economic Performance, Chicago: Rand McNally and Company, 1973, p. 367.

Exhibit 5-8

Project Cost as a Function of Time* (From Scherer)

General Model

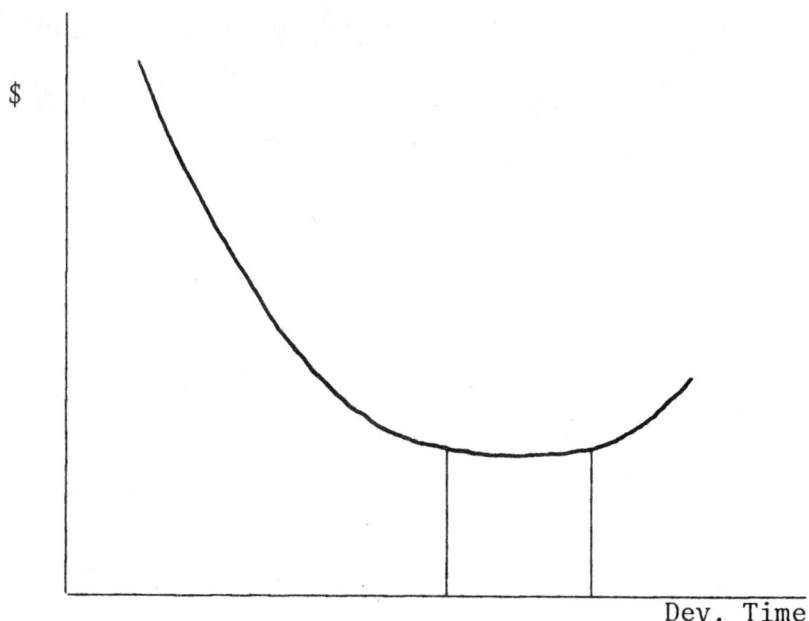

*F. M. Scherer, Industrial Market Structure and Economic Performance, Chicago: Rand McNally and Co., 1973.

Against the costs of accelerated development, a potential innovator must weigh the benefits of proceeding more rapidly. Early market entry usually reaps substantial profits. However, these profits must be weighed against the increased project cost.

On the other hand, long development times tend to increase costs as costs continue to accumulate, to say nothing of the increased costs due to inflation. (This has been particularly apparent with large time/cost overruns on government contracts.) It becomes prudent for management to plan projects so that the development time falls within a range of minimum costs commensurate with these other factors.

I felt that unsuccessful projects were using large amounts of manpower and allowed to run for periods of time that were almost as long as the successful projects. In order to measure this I examined the level of effort and elapsed time by project. By observing Exhibit 5-9, showing level of effort by project, it can be seen that the successful projects assigned, on the average, from 2 to 15 people to perform the R & D. The smaller projects, such as the one piece side panel for the car, or the electronic relay, or the fuel pump membrane had on the average 2 to 3 technologists working on the project. The larger projects, such as the dye for non shrink wool and the new front axle, averaged as many as 10 and 15 personnel assigned.

For the unsuccessful projects, the personnel assigned varied from a low of 2 technologists to a high of only 5. The mechanical dialing switch and the polarized headlights each averaged 2 technologists, while the print by chemical reactor, reactor dye for wool, and the ticket reader averaged 5 personnel.

Exhibit 5-9 shows that the elapsed time for the successful innovations varies from two to seven years, and quite similarly, the elapsed time for the

Exhibit 5-9

Manpower Assigned and Elapsed Time for Twenty-Two West German Innovations

Successful Project	Avg. Manpower Assigned per year	Elapsed Time in Years	Unsuccessful Project	Avg. Manpower Assigned per year	Elapsed Time in Years
Shock Absorbent bumper	8	3	Car door spring	3	1
Electronic relay	3	3	Mech. dialing switch	2	6
Modified Turbo Engine	6	2	Friction gear box	3	1
Dye/Non-shrink wool	10	7	Motorcycle st. motor	3	2
Super heat steam process	4	6	Print by chem. reaction	5	7
New front axle	15	4	Auto mech. shift	5	3
Improved tv circuit	4	3	Non-friction brake	3	5
One pc. side panel for cars	2	2	Reactive Dye/Wool	5	7
Fuel pump membrane	3	2	Ticket reader	5	3
Ignition Discharge System	6	5	Plastic milk bottle	4	5
Thin Film Circuits	5	7	Polarized headlights	2	4

unsuccessful projects varies from one to seven years.

By observing Exhibit 5-10 showing the summary level of effort, some interesting observations can be seen. The hypothesis is that significant amounts of manpower and long periods of elapsed time are being expended in unsuccessful projects. The average number of men assigned to the successful projects exceeds the average number of men assigned to the unsuccessful projects. The differences in the means are indicative of a direction, but the differences are not statistically significant, and it is, therefore, possible that these results occurred by chance.

The successful projects show an average assignment of six men, while the unsuccessful projects show an average assignment of 3.6 men. Let us put those facts aside for a moment while we examine the elapsed time, and then we will return to this.

Exhibit 5-10 shows that the total number of years, on the average, is identical at four years between the successful and unsuccessful innovations.[7] By examining the standard deviation, it can be seen that the unsuccessful projects show a slightly larger standard deviation than the successful ones. This could be expected since some unsuccessful projects are stopped after one year, although some go as long as seven years. However, the results are not statistically significant and this could have occurred by chance.

Now, let us see what all this means. One could surmise that if the unsuccessful projects had more personnel assigned during the same time periods, then perhaps these projects would have been successful. However, as alluring as that argument first appears, it does not seem to be a valid explanation as one examines the particular projects. For example, the polarized headlights would not have been successful if more personnel were assigned, and similarly

Exhibit 5-10

Level of Effort Summary

		Men Assigned	Elapsed Time in Years
Success	Mean	6	4
	Std. Dev.	3.79	1.95
Failure	Mean	3.6	4
	Std. Dev.	1.20	2.19

with the inflatable plastic milk bottle, the car door spring, etc.

Then what does one make of these findings? I propose that as innovation is taking place, the work performed during the period between the origin of the idea and the final market introduction does not exist in a vacuum. Information does start to enter the organizations and signals of success or failure start to appear. These early warning signs tend to encourage an organization to gamble with large personnel assignments or start them worrying and keeping the numbers low.

In addition, higher numbers of men are assigned toward the end of a project. Therefore, the mean number of men on the successful projects will be larger. The unsuccessful projects are terminated before that final heavy manpower commitment. Unfortunately, the record-keeping by project did not permit me to measure the exact manpower per year which often went back several years. The average manpower assignments were available. The increased manpower assignments toward the end of the projects are therefore a speculation on my part.

When faced with negative information, the firm has great difficulty with stopping the project but allows it to "limp along." That is why we see unsuccessful projects with elapsed times for as long as five, six, and seven years.

Most corporate officials feel that their R & D policies are based on relatively small, but constantly enriched product lines. The average R & D project is not long term; as a matter of fact, previous studies as well as this will show a mean of approximately four years from invention to innovation. These projects would generally be composed of incremental steps forward.

There are, of course, exceptions to the rule in which corporations embark upon major research projects of long time durations. Of necessity, these are performed by the larger corporations with resources sufficient to support such ventures.* As costs continue to rise, it is expected that the long term projects will have to continue to give way to the smaller changes. The drawbacks to the long run projects can really be summarized under three major categories:

a) Increasing costs often make very long run projects impossible;

b) The rate of technological and social change is so rapid that a span of many years from project start to finish will often result in a product or process that is no longer desired; and

c) The motivational feedback loops for the engineers, scientists, and other technologists become so long that it becomes difficult to obtain the commitment so necessary for a successful project. This is a particular problem in some industries where the projects are, of necessity, relatively long. This can be partially overcome with subsolutions along the way to a final solution. However, the completion and appraisal of results is an important part of the researcher's rewards, so longer run projects are faced with this additional difficulty.

---

*In a previous study that I did in 1970, I found average project times for Fortune "500" at 3.3 years and all others at 2.1 years. (A. Gerstenfeld, <u>Effective Management of Research and Development</u>, Addison Wesley, 1970.

The largest U.S. national R & D project in recent years was, of course, the ten year NASA program to get a man safely on the moon and return. It is now generally agreed that one of the most significant contributions of that program was toward the management of technology rather than the accomplishment of the feat itself.

Let us see if we can combine and generalize the findings on level of effort with the previously reported information on technology push. One would argue that since nine out of the eleven unsuccessful projects were the result of technology push, it perhaps becomes difficult to stop these projects. The ideas originate within the organization, and the individuals become committed. As the negative information is received, it either becomes distorted or blocked sufficiently so that the organization does not stop the project.8

In general, as the demand pull information is being received, it tends to support the firm's decision to embark upon the project, and the firm, feeling confident of the outcome, assigns large numbers of technical personnel. The signals from the outside encourage the organization, and larger efforts are made over the same periods.

Returning again to the unsuccessful projects, one observes an average elapsed time of four years and can argue that organizational inertia is still another reason that these projects are allowed to last for so long. An organization is willing to decrease the manpower assigned to the projects but appears unwilling to make the more difficult decision to stop the project. Clearly, with the benefit of hindsight, the organizations now realize that the danger signals were utterly apparent. The organization chose not to listen until vast investments of technical energy were exerted.

The data supply evidence for the prudent manager to exert tighter project control and to install early warning systems. It is obvious that the sooner one can detect a potentially unsuccessful project, the sooner the firm can stop work and direct its resources in a more constructive direction.

Perhaps the high project time for project failures results from companies putting in continuing effort as difficulties appear in the hopes of resolving those difficulties. For clearly it is a difficult decision for management to stop a project and perhaps more difficult than the decision to embark upon a project. One might emphasize to the decision-maker (be he in industry or public sector monitoring industry) that more concern should be directed to the decision for continuing. One should emphasize the importance of regular review periods and early warning systems so that no significant period of time may be allowed to elapse without a regular reporting procedure.

The projects that I investigated were funded by the firms themselves. It would be interesting to know whether government supported unsuccessful projects are drawn out longer than those financed by the firms themselves. There are two untested hypotheses for this:

- With government financing the projects are often more risky.

- Though bureaucrats are no more adverse to admit failures than people in industries, there is less pressure upon government officials to show results for the money they have allocated.

As mentioned briefly earlier, most companies stated that they had many more unsuccessful projects than successful. It behooves both the company and the country to take all steps for early detection of these situations. Perhaps these data will emphasize the importance for the manager to be constantly alert for early warning signs of project failures.

Process vs. Product Innovation

The problem of whether to invest scarce research and development resources into new products or into new processes poses a very important decision for management. It has been argued that process innovation trails product innovation for those cases where volume warrants it.* The emphasis for organizations today is mainly on new product development. The enormous research effort that is going on for new products exists because product life cycles are getting shorter, and it becomes necessary for the successful firm to continually develop new products.

It is also possible for the firm to adopt a policy of becoming a "follower" and simply imitate the products that the other firms develop. For some small firms this is sometimes the only available policy considering the high cost of research. However, this policy has associated high risks that are becoming continually higher. For as product life cycles become shorter, it becomes very difficult for the follower firm to bring the product to market in time for the profitable portion of the life cycles.[9] Obviously, patents also play a role in this case since the follower firm may be legally precluded from manufacturing the product due to the patent protection of the initiating firm. The patent issue is discussed in some detail in Chapter 3.

The counter argument which supports process innovation is, of course, the high cost of labor.

---

*W.J. Abernathy and Wayne, Harvard Business Review, Sept.-Oct. 1974.

Referring to Exhibit 5-1, item E, the super-heated steam process is an innovation as a direct result of high labor costs. The project manager stated the following:

> In the continuous process, the fixing time is the most important factor because it determines the production capacity. Short fixation times enable high fixation speeds and thus an economic production.[10]

In this case there was a strong demand for higher speed production and less manual labor. As labor costs increase, the incentive for process innovations, of course, increases. The operation described used to have manual workers carry the material from a rack after being coated with dye, and then roll these racks into large ovens. With low labor costs there was little impetus to change this procedure.

As labor costs rose, there was increased pressure from management on engineering to develop a continuous drying and fixing operation so that cloths would no longer have to be placed by hand on to racks and rolled into ovens. Development started in 1967 that resulted in a new process in 1971 that is now used by the company and is being sold to other companies as a major process improvement in the chemical-textile industry.

Another example of a process innovation recently initiated due to high labor costs is the removal of a soldering operation between two pieces of material placed together forming an auto front panel (Exhibit 5-1, item H). The filing and smoothing process became so costly that the company made the design investment to a one piece construction in spite of the initially high "one-time" dye costs. As labor costs rose and this company was placed in the position of having to be more competitive, the pressure mounted for process improvements.

There is an additional factor tipping the balance toward process innovation, namely inflation. A study by Poensgen and Straub* shows that inflation affects working capital more strongly than it affects fixed assets. Since product innovations normally require substantial increases in working capital (R & D plus marketing expenses), then the higher the rate of inflation, the more pronounced the shift toward processes innovation should be. The summary of Poensgen and Straub's findings regarding the effects of inflation on the product-process decision are as follows:

. With respect to process innovations, which result in fixed assets, the inflation causes today's investments to be paid off in future years with cheaper money.

. With respect to product innovations, the working capital necessary to produce these new products will continually increase as the inflation increases.

. Given a high rate of inflation, long-lived assets are far less affected than short-lived ones.

Therefore, we might continue to expect to see constant recognition of the importance of process innovation assuming the quantities warrant such investments. In actuality, the firm expends some costs on each. Those product lines that have high enough quantities, high labor costs, and promising futures often warrant process innovation.

---

*O.H. Poensgen and H. Straub, "Inflation und Investitionsentscheidung," Zeitschrift fur Betriebswirtschaft, Vol. 44, Dec. 1974, pp. 785-810.

On the other hand, the short product life and the large potential profits in new products influences the firm to concentrate in that direction.

These facts cannot be taken in isolation for the competition is always present and may affect the decision in either direction.[11]

Tax and Patent Considerations

The tax and patent policies of a country are often referred to as incentives for innovation. We shall first consider the question of tax policy as an incentive for innovation for West Germany. There are not a great number of specific tax incentives in West Germany and so the tax policy assumes a relatively minor role. The corporate taxes are indeed high and almost identical to the United States.

General R & D expenses are to a large extent tax-deductable. These sorts of expenses, such as salaries for R & D personnel, are deducted under the rule for expenses associated with basic scientific or technological research. Tax deductions for investment goods for R & D expired on 31 December 1974.

The driving force is generally competition, which was repeated in my interviews over and over again and can be seen by analyzing the current innovations. There appears to be no substitute for competitive spirit.

In response to tax considerations for the 11 successful and 11 unsuccessful innovations that I examined, I did not receive one response that indicated that the particular innovation was affected by tax policy. I think it is, therefore, safe to say that tax policy as an incentive for innovation in West Germany may serve as a mild overall incentive affecting the climate but certainly not a direct incentive for specific corporate projects.

For the successful innovations, 8 out of 11 had patent considerations, and 10 of the 11 unsuccessful innovations had patent considerations.[12] In many cases, the patent was not on the entire innovation but on portions of the innovation. For example, while developing the shock absorbent auto bumper, the company did patent portions of it. This was also the case for the development of the electronic relay, the turbo engine modification, etc. There were only a few of the innovations where there was no patented technology used and they were the super-heated steam process, the one-piece side panel for cars, and the motorcycle straight cylinder motor.

In Germany, the aim is to describe the invention in the patent claims as concisely as possible using technical terms of a general nature wherever possible. At the same time, the interpretation of the patent claims and consequently the definition of the protection thereby granted does not solely rely on the wording of the claims. For example, if the state of the art permits it, one is even inclined to interpret the patent claims in a very broad sense. Thus, it is not possible to legally circumvent a German patent merely by making use of the general inventive idea, omitting on the other hand say an inferior feature of the German patent claim. Considered from this angle, it is relatively difficult to go around a German patent.

The protection afforded under the U.S. Patent Law, in contrast, is more specific; this is reflected already in the wording of the U.S. patent claims which as a rule comprise more technical features than corresponding German claims. To circumvent a U.S. patent legally, it is often sufficient to omit one or two features that are defined in the U.S. patent claim.

Having considered a series of successful and unsuccessful innovations from many aspects, we now move on to an examination of a series of new innovations currently in work and not yet introduced to the market. Clearly, some will be successful and some will fail; but all will be influenced by the social forces of this decade.

Selected Readings

Abernathy, W. J. and Kenneth Wayne, "Limits of the Learning Curve," *Harvard Business Review*, Sept.-Oct. 1974.

Allen, T. J., "Communications in the Research and Development Laboratory," *Technology Review*, Vol. 70, Oct.-Nov., 1967.

Allen, T. J., "Communications networks in R & D Laboratories," *R & D Management*, 1, 1970.

Cyert, R. M. and J. G. March, *A Behavioral Theory of the Firm*, Englewood Cliffs, New Jersey: Prentice-Hall, 1963.

Freeman, C., *Science and Technology in Economic Growth*, (B. R. Williams, Editor), New York: MacMillan, 1971.

Gerstenfeld, A., "A Study of Successful Projects, Unsuccessful Projects, and Projects in Process in West Germany," *IEEE Transactions on Engineering Management*, Aug. 1976.

Lorsch, Jay W. and Paul R. Lawrence, "Organizing for Product Innovation," *Harvard Business Review*, Jan.-Feb. 1965.

Schmookler, Jacob, *Invention and Economic Growth*, Cambridge, Mass.: Harvard University Press, 1966.

Utterback, James M., "Innovation in Industry and the Diffusion of Technology," *Science*, Vol. 183, 1974.

CHAPTER 6

Social Forces Effects Upon Innovation

> I think that we can expect the rates of technological progress to be quite similar to those in the past. There will be all sorts of things now not available--maybe some not even dreamed of--that will be coming out on the market in the year 2000 to the benefit of the consumer. The pressure for improvement of social conditions--in its environmental aspects, safety aspects, and so forth--will be assuming an even bigger role in our economic life.*

The purpose of this chapter is to consider current projects in West Germany that may result in future innovations. From the 32 innovations studied, the 22 successes or failures were discussed in the previous chapter. We now turn to an examination of the 10 projects observed while in work. These projects are all in various stages of completion, but all are past the invention or origin point and are in the problem solving phase prior to the point of introduction to the market.

Third Generation Engineers and Scientists

Before reporting on the in-process projects observed in West Germany, I would like to first briefly consider the evolution in this century of

---

*Kenneth Arrow, "Capitalism and Society," <u>Business and Society Review/Innovation</u>, No. 10, Summer 1974.

West German engineers and scientists. Exhibit 6-1 shows three generations of technological growth. During the early 1900's, and until about 1950, the major emphasis of scientists and engineers was on technical accomplishments. Certainly two world wars influenced this with major emphasis then placed on performance. Of course the element of cost was present, but the degree of emphasis was clearly on technical performance.

Exhibit 6-1

Three Step Change in West German

Technological Thinking

1st generation	2nd generation	3rd generation
Technical	Technical and Economic	Technical Economic, & Social
1900-1950	1950-1970	1970-?

A shift took place in the early 1950's which I call second generation technical personnel. The emphasis remained on the technical, but the economic factors now took on increased importance. Engineers and scientists became much more cognizant of the cost implications, and project innovations responded to costs of labor and material.

Exhibit 6-1 shows that this period has lasted

until about the beginning of the 1970's. It is at this time that we move to third generation technological thinking. We retain the technical and economic emphasis, but now add the important social implications. For it was only a few scant years ago that ecology was not a word in our everyday vocabulary. The concepts of pollution and energy were of little concern, for the feeling was that our planet had unlimited resources.*

Social change is clearly interwoven with technological change. Technology affects the social system in a very significant way. The birth control pill is a good example of a technology clearly affecting family size in many countries. Similarly, the social changes affect technology; for example, as people wanted a higher speed of travel, the pressure mounted for faster planes. Now the priorities are more closely linked to concerns about urban transportation. It has been said that all technologists should be more careful about what they are doing since they are really affecting the social system as a whole in a significant way.

Motivating Forces for Projects in Work

Since we are examining innovation as a result of industrial projects, the overriding drive is, of course, profit. When I asked the basic question, "Why

---

*Studies such as Meadows, "Limits to Growth" served the important function of alerting many to possible dangers. Almost all of the engineers and scientists interviewed, when asked about future innovation, cited "Limits to Growth" and stated that we must continue to innovate but smarter and safer. The New Club of Rome Report, "Mankind at the Turning Point," more accurately stresses the importance of selective innovation.

are you engaged in this project?" the response was
always "because we are in business, and because we
must make a profit." However, as I dug for a deeper
level of understanding, the word competition started
to appear. The common response was that "we must be
engaged in this project because our competition will
do it if we don't."

Both of those levels of response are of interest
and makes one feel that our Western, profit oriented,
competitive society will continue to innovate.
However, in helping us to better understand the direction of innovation, one must unravel the facts to an
even deeper level of understanding. We shall consider
profit and competition as a given, and from these, let
us see if we can observe patterns to the sampled
innovation projects that are in work.

Exhibit 6-2 separates out the major motivating
forces for this sample. For example, the anti-skid
device which is currently in work is in response to
an increasing emphasis on automobile safety. (This
particular project is discussed in more detail later
in the chapter.) Item C in Exhibit 6-2 is being
developed in response to two major forces. The new
process is to reduce a labor operation from two
minutes to ten seconds, and in addition, eliminate an
acid spray that is the result of the current procedure.

By observing Exhibit 6-2, I think one would have
difficulty finding an overriding pattern but, indeed,
one could conclude that projects currently in work,
in this sample, are as a result of a multitude of
forces. The forces are complex and projects are
often the result of several forces.

Of the ten projects I observed, the high cost
of labor was impetus for some. For others the impetus
was the high cost of material. In addition, several
projects are resulting from an emphasis on increased

Exhibit 6-2 (continued on next page)

Major Market Forces for Decision to Undertake
Eleven Current West German Projects

Project Description	Labor Costs	Nat'l Costs	Safety	Pollution	Energy	Misc.
A. Anti skid device for cars			X			
B. New dye material which adds moisture first to eliminate bad smells in factory				X		
C. New process for putting color dye on material—Time reduced from 2 min. to 10 seconds, plus no acid sprayed in air	X			X		
D. Faster telephone switching device						X^1
E. Pressurized engine with different injection system						X^2

Exhibit 6-2
(cont'd)

Project Description	Labor Cost	Mat'l Costs	Safety	Pollution	Energy	Misc.
F. Two system braking for trucks			X			
G. Home heat pump as an alternate energy source					X	
H. Drinking water filter				X		
I. One piece auto hood of plastic	X	X				
J. Development of new extra large motorcycle with sidecar for general transportation in USSR					X	

[1] Larger population using telephones, requiring faster switching service.
[2] Very high speed car for use on roads such as no speed limit autobahns.

safety, reduced pollution, and reduced energy utilization.

As one observes the projects in work, it is an opportunity to peer into the world ahead, which I am afraid shows much confusion. For example, item A in Exhibit 6-2 is an anti-skid device to increase auto safety. Simultaneous to this development, item E in Exhibit 6-2 shows a pressurized engine that can greatly increase the speed of cars. Thus, as an example of the schizophrenic world we are moving toward, one sees emphasis on auto safety and simultaneous emphasis on the development of still faster cars that are the very antithesis of safety. And they will probably both be successful. If one must have an explanation, it is of course that organizations cater to diffuse markets. As safety gains in importance to many potential buyers, so the thrill of high speed cars will gain in importance to others.

Example A

In trying to better understand innovation it is helpful to realize that the innovations at one date are the result of all other related innovations up to that date, plus the sum of the motivating forces at that particular time.*

Let us now trace through Item A from Exhibit 6-2, the anti-skid device. The anti-skid device is being developed in response to the forces of safety.

---

*To present this concept mathematically we would say that innovation at time k is a function of all other innovations, related to that item, up to time k, and a function of all the forces acting a time k:

$$I_k = f(I_{k-1}, I_{k-2}, \ldots, I_0, F_{1,k}, F_{2,k}, \ldots, F_{m,k})$$

where I = innovation and F = motivating force.

I prefer to use the word force rather than
government regulation, since at this time we are not
concerned with the riddle of whether social pressure
causes governments to react, or government influences
social pressure. The anti-skid device is not currently
a legislative requirement, and in the future it may
or may not become one.

The objective is to develop a product that
would add safety to the automobile and accomplish
this result at a reasonable cost. This is expected
to be a major automotive innovation.

In 1968, a brake company initiated the idea for
the design of a low cost anti-skid device to be sold
to automotive companies.* The car, long the mechanical engineer's undisputed masterpiece, has given way
to electronics in the interest of safety and pollution
control. The brake company felt that the
project had merit and so allocated funds for the
beginning of a significant development effort. This
project is a technology push and is consistent since
this will be a very major innovation if successful.
The sequence for the anti-skid device is shown in
Exhibit 6-3.

After one year of work, in 1969, with the project somewhat underway, the brake company presented their ideas to several automotive companies in
order to gain some feedback. They received positive
information and then set upon a major project that is
in work now with an estimated completion date of 1978.
This is, therefore, a ten year company funded project

---

*Part of this company's motivation is that
patents on certain products which have been a major
contributor to corporate profits are running out. In
anticipation of this, a major R & D program is underway. The effect of patent life on innovation is
discussed in Chapter 3.

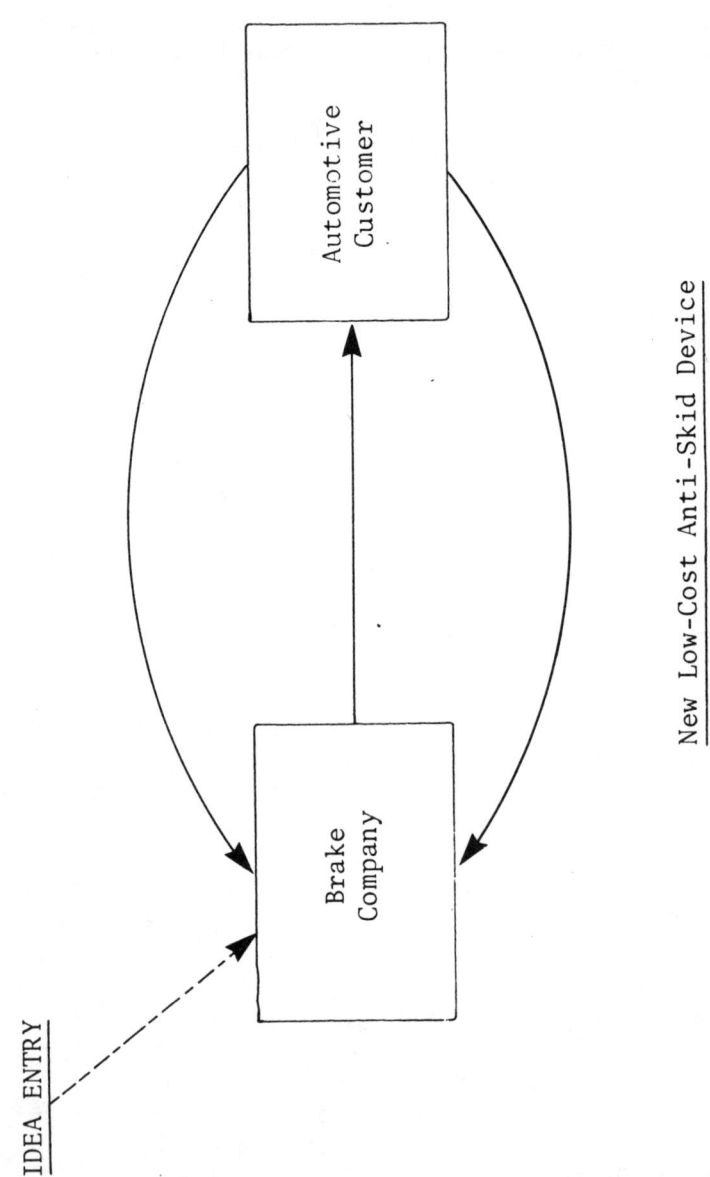

New Low-Cost Anti-Skid Device

Exhibit 6-3

representing a substantial investment and risk.*

The concept of the innovation being a function of all other related innovations is particularly clear in this case. Let

I_{k-3} be the transistor

I_{k-2} be the printed circuit boards

and I_{k-1} be the solid state chips

The development of the transistor paved the way for the beginning of this project. By using transistors and electronic controls the braking action could be achieved. These transistors would then be mounted under the hood along with the associated circuitry. This would allow the circuit to be activated and then cut off at a rate of approximately four times per second. As a result, each wheel would have the brakes applied and released at a rate of four times per second, which is immeasurably faster than a human could respond. The car would come to a rapid halt but without the locking action which often causes skidding.

The innovation of the solid state chip meant that the equivalent of the transistors and the hundreds of components could be placed in one piece, smaller than one inch by one inch at low cost. Without the benefit of low cost solid state chips, the device as originally conceived would add too much to

---

*This presents an interesting example of Professor Galbraith's contention of the power of a few large buyers. In this case, the power of a few large automobile buyers is causing the supplying companies to innovate. John K. Galbraith, <u>American Capitalism: The Concept of Countervailing Power</u>, Cambridge: Houghton Mifflin, 1952, ch. 9.

the price of the car. With the chips, the available technology could be coupled with the forces for increased automobile safety at reasonable costs.

The project is presently at about 50% completion with 1978 as the target date. During the development of the anti-skid device, the brake company continues its interaction with the automotive companies as prospective customers. The automotive companies are free to contribute ideas to the design, although, in actuality, the electronic project engineers are at least two years ahead of the automotive electronic engineers, and the automotive inputs are minimal.

In summary, what do we learn from this current project case?

- We observe a company willing to gamble on a ten year project.

- We see the workings of third generation engineers--technical, economic, and social.

- We see clearly the cumulative effects of technology.

- We see the technological growth coupled with the motivating forces, and the combination resulting in a potential and major innovation.

Example B

In 1972, a major West German corporation embarked upon a six year project that is shown as item H in Exhibit 6-2, a drinking water filter. This project is at about the 50% point, and, again, illustrates a project as a function of forces and all other related innovations to date.

This particular project was launched as the drinking water in many parts of the world has become increasingly contaminated. It is true, of course, that major efforts are underway to reduce the pollution in major world rivers. Nevertheless, the water drawn from the ground near these rivers will continue to be a problem. As a result of the need for a drinking water filter, the company embarked upon the research project.

The accumulated innovations that may lead to the filter are new synthetic fibers as well as low cost, highly reliable small pumps that would force the liquid through the fibers. The small filters might be installed in each home and be used not only to improve drinking water, but also used for washing machines to convert hard water to soft.

For both of these projects it becomes particularly apparent how sensitive the technological and market antennae must be. Changing market forces must be sensed and as new technologies develop, information about them must be assimilated by the organization. It is the integration of the new technologies and market forces that combine to provide the firm the possibility of embarking upon successful projects.

The channels by which this information is transferred to the firm leads into the entire issue of diffusion, the subject of Chapter 7. I have ever increasing respect for the importance of accurate and fast information transfer. Information transfer is best performed on a person-to-person basis. The importance of bringing new people into the organization with their new ideas remains as a significant goal. It is equally important that the existing members of the organization go beyond the organizational boundaries and meet with professional organizations, customers, and suppliers.

We have briefly considered the third generation engineers and scientists, the broad array of motivating forces for projects in work, and several examples combining the social forces and the cumulative technology.

Chapter 6

Suggested Readings

Arrow, K., "Capitalism and Society," Business and Society Review/Innovation, No. 10, Summer 1974.

Bell, Daniel, The Coming of Post-Industrial Society, New York: Basic Books, 1973.

Bell, Daniel, The Cultural Contradictions of Capitalism, New York: Basic Books, 1976.

Deane, R. H., "Minimizing Legal Liability for Unsafe Products," IEEE Transactions on Engineering Management, Vol. EM-22, No. 4, Nov. 1975.

Mesarovic, M. and E. Pestel, Mankind at the Turning Point, New York: Dutton, 1974.

CHAPTER 7

Technological Diffusion as Part of the Innovative Process

Diffusion is the gradual mixing of the molecules of two or more substances, or the angular redistribution by a scattering, reflecting, or refracting system. . . .*

The physical analogy of the mixing of the molecules and the subsequent redistribution process can be applied to innovation and technology. A firm's technological system receives information from the outside, converts this information, and in turn delivers information. The technological system is both the recipient of diffusion and the generator of diffusion. An R & D project involves the searching and obtaining of outside information, converting of this information, and delivering of information in the form of a design. The problems of diffusion are critical, and, therefore, affect both the receiving and the sending of information. In this chapter, we shall first consider the firm as an information processor, then diffusion of basic research, and the final section shall be on the West German universities and their roles in diffusion.

Diffusion: The Firm's Technological System as an Information Processor

Companies' technologists, engineers and scientists, must <u>receive</u> information from the outside, <u>process</u> the information, and then, in turn, <u>supply</u>

---

*<u>American Heritage Dictionary</u>, Houghton Mifflin, 1970.

information to the outside.[1] In general, if any of these channels are lacking or poorly functioning, a firm cannot be effective. For example, when a firm subcontracts portions of its R&D, the subcontractor firm performs its own research and delivers information to the contracting firm in the form of drawings or specifications. If the information comes in distorted, then the contracting firm will be poorly functioning.

It has been shown that information is usually brought into the firm by a few individuals termed, "technological gatekeepers."* These individuals have more extensive contact than the others do with colleagues outside the firm. The technological gatekeepers usually have more familiarity with the technical literature. These persons are usually used as internal consultants. It has also been shown that the technological gatekeepers are generally in close communication with others in the organization who share these characteristics.

It is equally important for the firm to know how to use these sources of information, products, and services.[2] In this section, we shall focus on the outside assistance that the firms studied in West Germany use for innovation. Exhibit 7-1 shows the general process. The exhibit shows some of the major sources of information.[3] Our investigation is limited to four outside sources as follows:

a) government agencies and laboratories

b) universities and consultants

---

*Allen, T.J., Technology Review 70, Oct.-Nov. 1967; Allen, T.J. R&D Management, 1, 14 Oct. 1970.

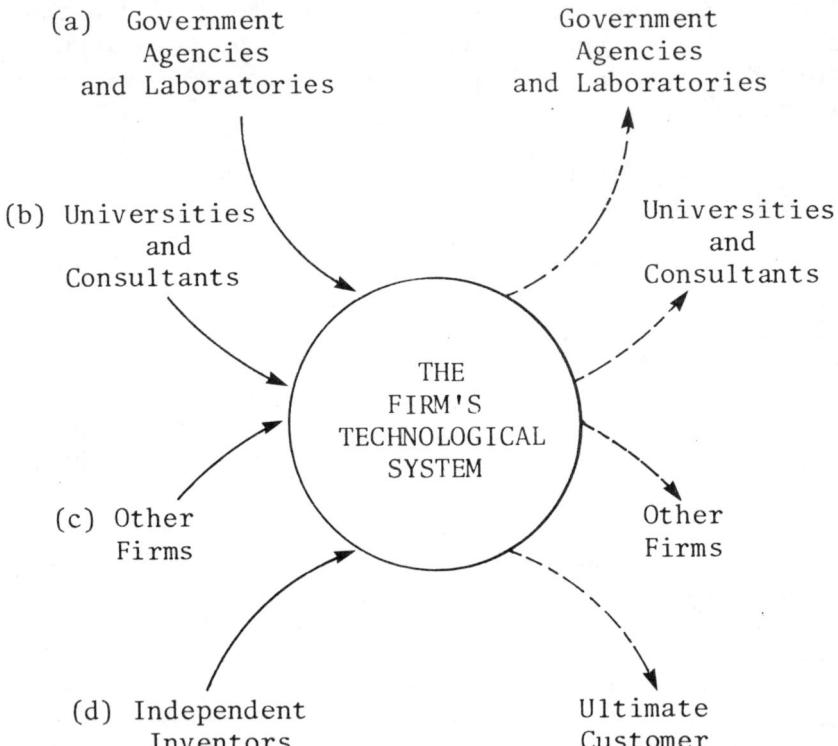

The Firm's Technological System
as an Information Processor:

A General Model

Exhibit 7-1

c) other firms (subcontractors)

d) independent inventors

a) Government Agencies and Laboratories

The government's contribution in U.S. and West Germany in stimulating innovation is very important. We observed in Chapter 1 that this contribution is both direct and indirect. The direct mechanisms include government and industry, and government contributions to industrial joint research associations. The indirect government mechanisms to stimulate innovation consist of tax incentives, regulations, patent laws, etc.

In addition to these funding mechanisms, the government can also serve as an information source to firms. I investigated this aspect of the diffusion system for 11 successful and 11 unsuccessful projects in West Germany. The results are shown in Exhibits 7-2A and 7-2B.

None of the 11 successful innovations used the help of a government agency or laboratory and none of the 11 unsuccessful innovations used government assistance. There are some precautions to be taken, however, when interpreting this data. For instance, there may have been some basic government funded investigations performed several stages earlier that aided in the development of the particular innovation. However, in terms of a direct information channel for the innovations within this study, the link to the government was not visible.

b) Universities and Consultants

The role of universities and consultants in West Germany was examined as shown in Exhibits 7-2A and

Exhibit 7-2A

Use of Outside Information Sources for Eleven
West German Successful Innovations

Successful Innovations	a/ Govt	b/ Univ. & Consult- ants	c/ Other firms	d/ Independent Inventors
1. Shock absorbent auto bumper	No	No	Yes	No
2. Electronic Relay	No	No	Yes	No
3. Turbo Motor modification	No	No	Yes	No
4. Dye for non-shrinking wool	No	No	Yes	No
5. Super heated steam process	No	No	No	No
6. New front axle for auto	No	No	Yes	No
7. Improved TV Circuitry	No	No	Yes	No
8. Replacement of soldered joint	No	No	No	No
9. New fuel pump membrane	No	No	Yes	No
10. Ignition discharge system	No	No	Yes	No
11. Thin film circuits	No	No	Yes	No

Exhibit 7-2B

Use of Outside Information Sources for Eleven West German Unsuccessful Innovations

Successful Innovations	a Govt	b Univ. & Consult- ants	c Other firms	d Independent Inventors
1. Cam action spring for car door	No	No	No	No
2. Mech. switch for Tel. dial	No	No	Yes	No
3. Friction gear box for auto	No	No	No	Yes
4. Motorcycle straight cylinder motor	No	No	No	No
5. New process to print by chemical reaction	No	No	No	No
6. Auto mech. shift	No	No	No	No
7. Non-friction brake	No	No	No	Yes
8. Reactive dye for wool	No	No	No	No
9. Automatic ticket reader	No	No	No	No
10. Inflatable plastic milk bottle	No	No	No	No
11. Polarized headlights	No	No	No	No

7-2B, and was found to be seldom used. Only one of the 22 innovations used the university or consultant as a direct information channel. For the technical issues, the companies generally felt that they were ahead of the industrial consultants.

There is generally little direct linkage between the firm and the university in West Germany. Although the firms seldom use the universities as a direct information channel, the universities are linked to industry in less direct ways. This is discussed in some detail later in the chapter. However, an example of direct technical assistance from a university to a firm involved the design of a turbo motor modification requiring testing in a large wind tunnel (Exhibit 7-2A, Item 3). For this work, the university facilities and faculty assistance was used. This use of university facilities is infrequent, although it is not unusual for a company to invite a university professor to speak to a technical group regarding some general principles.

c) Diffusion Between Firms[4]

Exhibits 7-2A and 7-2B show the participation of other firms for each of the innovations. Interestingly, it was found that successful innovations showed a greater use of other firms while most of the unsuccessful projects showed no use of other firms.

Exhibit 7-3 shows that, of the innovations examined, 9 of the 11 successful innovations used subcontractors. The reverse was true for the unsuccessful innovations. For these unsuccessful innovations, only 1 out of the 11 innovations used the assistance of other firms.

This brings us once again to the argument that information is being received by the organization during the problem solving phase. When the information is positive (and it appears as if the project will be successful) then the organization gains confidence in the project and enters into subcontract agreements. On the other hand, as the project starts to run into difficulties, the certainty is lacking and the company hesitates in entering into subcontracts.

Exhibit 7-3

The Relationships Between the Firms Use of Other Firms and Successful or Unsuccessful Innovations*

	Successful Innovations	Unsuccessful Innovations	Total
Firm used assistance from other firms (use of intercorporate linkages)	9 cases	1 case	10 cases
Firm did not use assistance from other firms (no use of intercorporate linkages)	2 cases	10 cases	12 cases
	11 cases	11 cases	22 cases

*The Fisher Exact Probability Test shows $p < .005$.

It is equally reasonable to assume that the unsuccessful projects have early indications of low probability of success and, as such, prefer a low profile. The expenditures in-house are usually not monitored as closely as outside separate contracts and so the firm prefers to stay away from subcontracts.

Another explanation for the high relationship between successful innovations and high use of other firms is that the technical expertise of subcontractors is necessary for today's complex projects. The individual firm is no longer able to have the necessary breadth essential for current technologies.[5] Firms become managers for technology with some in-house expertise, but the firms clearly need the subcontractor's specialties to augment the firm's own efforts.

As a final postulate to help explain the low level of subcontracting for unsuccessful projects, we can return once again to the concept of technology push. Chapter 5 showed that for unsuccessful innovations, technology push was often the origin of the project. The idea entered the firm from the technical group and so the technologists prefer to hold on to the work with a highly personal commitment. Certainly this phenomenon was true in the case of the cam activated spring where the company could only get the engineers off of the project by finally transferring them to other portions of the organization.

And so again we see that early warning signs are available to the prudent managers. Perhaps there is even a regenerative process where positive information leads to confidence, leads to subcontracting, leads to success, etc.

At any rate, as one views outside technical assistance for our sample, we see little direct interaction between company and government, university, or consultants. In successful projects we do, however,

find a strong technical relationship between firms--
strong intercorporate links.

d) Independent Inventor

The role of the independent outside inventor was also investigated during my interviews. The research efforts of today require teams (Chapter 2) and large accumulations of capital (Chapter 4). However, the invention portion of the innovative cycle is still relatively inexpensive so I felt it would be of value to examine today's role of the independent inventor. The results on this issue were very much as expected. Exhibits 7-2A and 7-2B show that of the 22 successful and unsuccessful innovations, only 2 came from outside inventors. Of the two, both were unsuccessful innovations.

During presentation of these findings at a leading German university (University of Saarland) it was suggested that perhaps this low use of outside inventions is due to the fact that the companies sampled were large companies. Their argument was that large companies can afford their own inventors and research groups so we might expect to find a low frequency of outside inventors. I certainly concur, and one can only conjecture what the results would be with smaller companies. My belief is that the use of independent inventors in Germany is very limited, indeed, although perhaps not quite as limited for the small companies as for the larger ones.

Mansfield states, ". . . independent inventors continue to produce a significant share of the important inventions, although their relative importance seems to have declined."* I found that the decline

---

*E. Mansfield, Technological Change, New York: W.W. Norton, 1971, p. 41.

is even more significant than suggested, although no company is willing to actually close the doors to the possibility of the use of outside invention.

As another issue regarding the use of outside inventors, it is important to consider one of the initial concepts discussed in Chapter 1. The invention is but a small part of the innovative process. As such it represents a small portion of the cost of innovation.

This fact could relate to the small use of outside inventors, because the firm must make the major investment as post invention. The firm hesitates to invest precious and limited R & D funds on someone else's work. The company generally has more confidence in its own inventions, and is less timid about additional expenditures. Perhaps this is rightly so since the "in-house" invention is better understood by the personnel and perhaps equally important is the strong commitment that is concomitant with one's own invention. Funds can then be invested in the more expensive portions of the innovative process.

The phrase that is used in Germany for the independent inventor is "knap sack" inventor. One patent attorney stated, "the knap sack inventor has a very small role."

It is not only the capital and time that is required for today's complex innovations, but it is the emphasis on cross disciplinary fields that is needed. The cross fertilization and mixture of specializations, so endemic to today's technology, does almost preclude the single inventor. For example, the anti-skid device (Chapter 6, Exhibit 6-2, Item A) is using three basic fields of expertise as shown below:

Theoretical Physicists	Hydraulics Engineer	Electronics Specialists
philosophy of designing an anti-skid device	puts idea into practice	circuitry for hydraulic valves

This cross fertilization will result in co-inventors. The team consisting of theoretical physicists, hydraulics engineers, and electronic specialists will pool their knowledge during the work on this project. As technology becomes increasingly complex, the utilization of an individual inventor as contrasted to trained teams becomes decreased.

Another particularly dramatic example of mixed teams was cited by a chemical manufacturer working on dyes for placing colors on non-shrink wool. For approximately three years of R & D, the team made little progress. The company then added a textile engineer to the team. Due to a new perspective, the team realized that they were going in the wrong direction. They changed direction in 1970, and within three years, the innovation was successfully presented to the public (See Chapter 5, Exhibit 5-1, Item D).

Returning to the subject of the role of the independent inventor, several interviewees gave specific cases where the company had been hurt financially through the use of outside independent inventors. One example is the friction gear box (Exhibit 7-2B, Item 3). In this case, the outside inventor claimed, when initially approached by the company, that the problems were 90% solved and the company only had to complete the last 10% before having a successful project. This analysis did not turn out to be true, and after many dollars and years, the company dropped the project and accepted its loss.

Another way of viewing the difficulties of the independent inventor is to use an electric circuit analogy. On one side of the circuit is the firm. On the other side of the circuit is the outside inventor. When linkage exists through a coil, there is often a high impedance mismatch. One can say that the outside inventor's relation to the organization is often a high impedance mismatch.

The concept of internal commitment also adds to the difficulty faced when trying to use outside inventions. The many steps that must be taken to convert the invention to an innovation must have the whole hearted support of the organization. It is too often that this is lacking when the invention is from the outside. In my investigation of the cases where the outside invention was used, and the project unsuccessful, I kept hearing, "it was not invented here." This has been referred to in other situations as the NIH, or "not-invented-here" syndrome which I am afraid is often used as the excuse for failure.

One cannot leave the subject of independent inventions in West Germany without citing the case of the Wankel engine. One of my interviewees had met with Wankel at least 50 times and could speak knowingly of his invention. It was explained that Mr. Wankel had devoted his life to this invention but that there were basic design difficulties. The Wankel engine was studied very systematically, but, of course, was presented at a bad time with the forces toward fuel economy making the product undesirable. The Wankel case was cited as one of the few inventions in which the inventor became rich and the companies lost money.

In spite of the difficulties in the use of independent inventions, the interest in the independent inventor remains alive. One might say that the flame is greatly diminished but not extinguished. Each company stated that the independent inventor is still a force that should not be discarded.

Having considered the firm's technological system as an information processor, let us briefly review the results of my interviews. Exhibit 7-4 summarizes the findings on the use of outside sources. The government laboratories and consultants were not used, the universities and consultants were unused (except on one occasion), the other firms were used extensively for the successful innovations, and the independent inventor was seldom used. The firm does not exist in isolation, the firm does use some outside sources, and it is the judicious use of such sources that can effect the success or failure of a project.

Diffusion: Basic Research

The diffusion of basic research usually takes place through professional meetings and published papers. In Chapter 1 it was shown that the industrial joint research associations in West Germany are effective in working on common and fairly basic industry problems (i.e., new auto anti-pollution concepts, new materials, etc.). As these research groups become involved with more applied research, competition sets in and the effectiveness of the diffusion is greatly diminished.

This pattern is true also on an international scale. For example, Euratom (European Atomic Energy Commission) was meant to combine European resources for nuclear research. The aim of Euratom was to create the conditions necessary for the speedy establishment and growth of nuclear industries by:

- promoting R & D and dissemination of technical information;

152

Exhibit 7-4

Summary Use of Outside Sources for
Twenty-Two West German Innovations

Projects	(N)	a Use of Govt	b Use of Univ. & Consult.	c Use of Other Firms	d Use of Indep. Inventor
Success	(11)	0	1	9	0
Failure	(11)	0	0	1	2
Totals	(22)	0	1	10	2

. facilitating investments and ventures;

. assuring a regular supply of ores and
  nuclear fuels; and

. creating a common market in material
  and equipment.

Euratom was largely a failure. While engaged in basic research, all went well, but as solutions developed and the research became more applied, difficulties arose. Individual national interests became paramount, and the program was forced to change directions.

The nuclear research centers have now moved back into more basic research in fields such as environmental, waste disposal, raw material, etc., and diffusion of information can be more readily accomplished.

In Europe, a new organization formed in November, 1974 known as the European Science Foundation (ESF) will coordinate international basic research.* West Germany and fifteen other countries will participate, nine of which are the Common Market countries. As a first step, basic research on accelerators, astronomy, archeology, theoretical math, and biology are areas which will be studied. Additional research will be added at a later date, including the fields of law, medical research, etc..

To aid in the diffusion of information, West Germany and the other countries will exchange scientists so that transfer of knowledge can take place

---

*Dr. Albert Strub, Science Advisor to the E.E.C. who was involved with the formation of the European Science Foundation, was helpful to me in our discussions concerning the new organization's goals and objectives.

beyond national boundaries.*[6] The ESF will also originate meetings to further transfer technology and fund joint projects, thereby serving as an important information channel for basic research. Included in the 16 countries are 47 member research organizations, each performing their own research. These organizations represent the efforts of 150,000 scientists, engineers, and other technical personnel with a total annual budget estimated at 1 to 2 billion dollars.

To aid in diffusion, the European Science Foundation recognized the importance of physical location.** (Note that the Route 128 studies cited nearness to M.I.T., Harvard, and Boston University as one of several important factors responsible for the area's success.) The location of Strasbourg (France) for the European Science Foundation was selected because it is near several scientific universities: University of Louis Pasteur, Strasbourg, France; Universities of Karlsruhe and Stuttgart, Germany; and University of Basel, Switzerland. Swift access to capitals and important centers of science in Europe is possible by train and airplane.

---

*The European Science Foundation statute lists as one of the principle objectives, "to assist the free flow of ideas and information." (European Science Foundation, Statute, Article II, Objects of the Foundation, i.c.)

**In my previous text on R&D, I emphasized the importance of physical layout on information flow. In that study it was internal (within the firm) but the findings are equally applicable for external (i.e., where a research center is located). A. Gerstenfeld, Effective Management of Research and Development, Reading, Mass.: Addison Wesley, 1970, p. 72; also see Allen, T.J. and Fusfeld, Technology Review, M.I.T., May 1976.

Diffusion: University-Firm's
 Technological Interface

The relation between the universities in West Germany and the firm's technological system is shown in Exhibit 7-5. The university and the firm are separated by a barrier as shown although the interdependence is present in less direct ways. As shown at the top of the diagram, the universities' work in basic research often provides an underpinning for the industrial research. The loop works in the other direction as well since the basic research at the university is often spurred on by the applied research at the company.

Considering the second from the top loop in Exhibit 7-5, it can be observed that university professors connect to the firm's technological system through basic principle lectures. Several firms have this procedure on a regular basis. This loop also is bidirectional since these lectures often provide an opportunity for the university faculty members to receive information from the firm's technological system.

Exhibit 7-5 shows a small arrow indicating direct consulting (usually in pharmaceuticals or chemicals and occasionally in other industries). This information channel also works both ways. The direct consulting relationship is less common in West Germany than in the United States and is largely due to large firm size and in-house capability in West German firms plus the more theoretical nature of West German universities.

The next to the bottom loop in Exhibit 7-5 shows the industrial research organizations providing an indirect link between the university and the firm's technological systems. The faculty do perform consulting functions for the industry as a whole, and information is exchanged, again showing a two-way

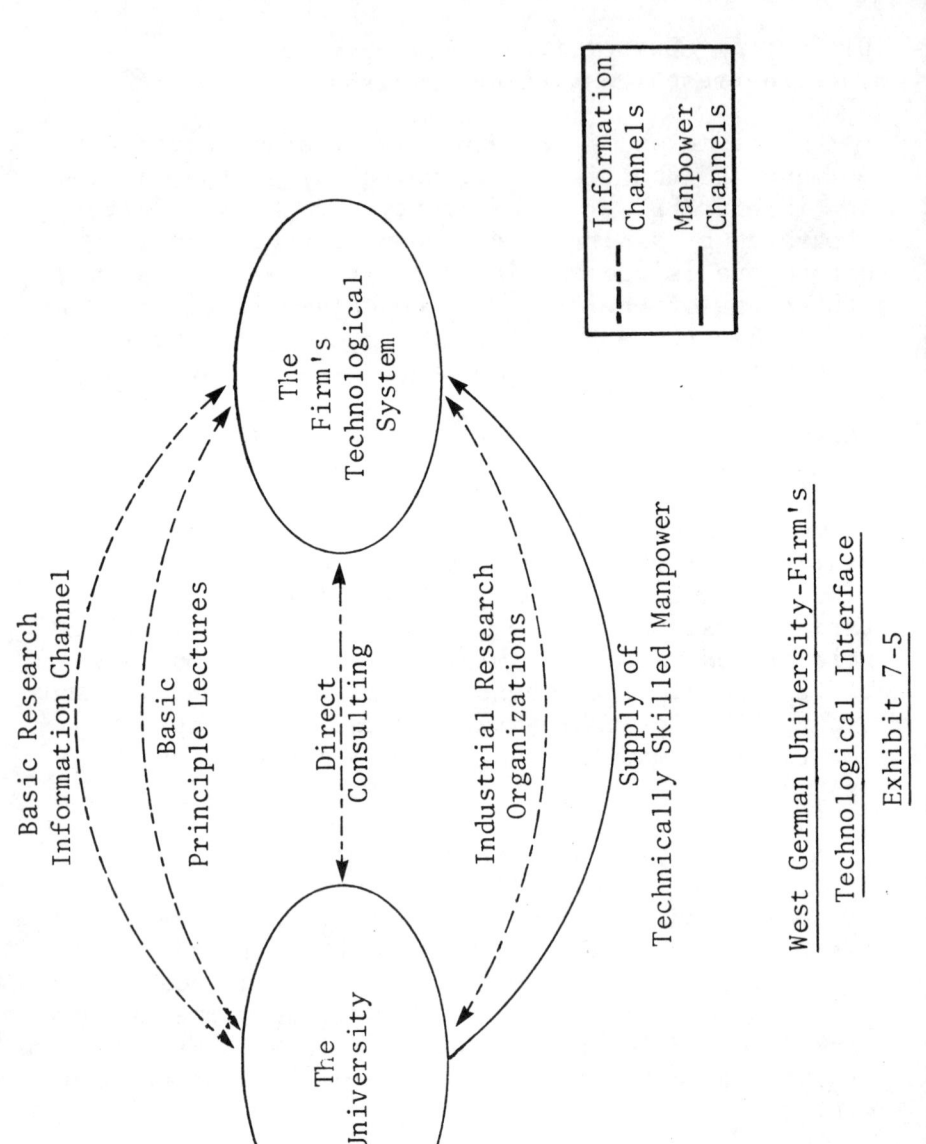

West German University-Firm's
Technological Interface
Exhibit 7-5

information channel. The firms participate indirectly since the research findings are shared among the members.

The final link of the university-firm relationship is the obvious responsibility of the university to supply skilled technical manpower to the firm. This supply is ever increasing with fairly standard output and little variation among universities. This relationship is one-way, i.e., from the university to the firm, since we do not see the manpower transfer from the firm to the university as we do in the United States. The information channels are therefore two-way while the manpower channel is unidirectional. We shall now consider each of these links between the university and the firm's technological system in more depth.

Basic research that is performed at the university often supplies necessary information to aid in the solution of the firm's applied research (top loop Exhibit 7-5). The West German policy is that a faculty member spends on the average 45% of his time in research, 45% of his time in teaching, and 10% in administration.

Within the West German universities, there is an equal commitment to create new knowledge (research), and pass along knowledge (teach). This shared commitment is being challenged at present and research is giving way to teaching due to ever increasing enrollments. (The enrollment issue is examined later in the chapter when considering the universities' role in supplying manpower.)

There has never been any quarrel over the government's firm commitment to support research. I was concerned about whether this close tie between government and university might be of damage to the university system. I addressed myself to this subject

with several of the West German faculty. They seemed
to feel that there were some dysfunctional aspects to
this system. These disadvantages were described as
largely bureaucratic. For example, the flexibility
that might exist in other situations is lost. If
money were allocated for one research assistant in
physics, then the professor could not lure two half-
time assistants for the same total pay. Or, if an
assistant professor were to be hired in electrical
engineering and it were specified that he was to have
certain qualifications, he could have many other
qualifications but unless he fit the tight formula,
he could not be admitted to the faculty.

The second loop in Exhibit 7-5 shows the interface between the university and the firm through basic
principle lectures. Industry in West Germany generally sees the university as a training ground for
broad principles. Many of the larger corporations
have regular monthly lectures given by University
professors. The topics are usually broadly related
to the firm's interests but not meant to solve specific problems. Notices are sent out in advance and
attendance is usually voluntary and quite large.
Needless to say, the information channel functions
both ways since such exchanges afford the opportunity
for the university faculty member to be updated on
the state of the art as practiced by industry.

The short direct link between the university
and the firm in Fig. 7-5 is in direct consulting.
Although the direct relation between industry and
university in West Germany is generally not common,
there are some exceptions. The principal exception is
in the field of pharmaceuticals and chemicals.
Pharmaceutical researchers and organic chemistry researchers in general have a very close interaction
between industry and university. University Ph.D.
students do work in industry, and professors act as
consultants, and the federation of chemical industries

maintains lists of what each professor is working on.

Figure 7-5 shows an information channel toward the bottom linking the university and the firm through the industrial research organizations. As shown in Chapter 1, there are 76 industrial joint research associations in West Germany in which most corporations are members and share in the costs along with the government. Research is performed by these associations for the benefit of all of their members. During this R & D, it is not unusual to enlist the assistance of university faculty.

Recently, a faculty and industrial research association joint effort took place on a new kitchen design. The faculty aided in time study and motion economy, and the association came up with standards benefitting all member firms. This collaboration among firms, and among firms and universities, illustrates several points. First, in this case, the firms cooperated on a non-basic, very applied problem. Secondly, there was interaction between industry and academic faculty.

Let us now examine the universities' role in supplying well-educated personnel (Bottom loop Exhibit 7-5). Until recently and by law, all students who wanted to attend a West German University were admitted assuming completion of preparatory schooling. There were no grade levels necessary to be admitted and there were no entrance examinations.

The system is now changing due to the rapid expansion of the university student applicants. The universities are offering very few new professorships, to go along with increased enrollments, and the requirements for entrance are now limiting the accessability of the opportunity for higher learning. Students are being turned away and for the sciences and engineering this is clearly considered a step forward by some academics. The student may apply again the following year, and in fact his priority will increase. Nevertheless the classes remain large, and there is little personal interaction between professor and student. As entrance requirements continue to be enforced, the average competence of students is expected to increase although it is expected that a system of mass education will remain the norm.

The percentage of grade school students that continue their education in high school, and of high school students that want to get a college degree has increased continually in West Germany, and is expected to increase further. Exhibit 7-6 shows the number of students in West German Universities planned for the years 1978 through 1990. These figures are based on past experience and the assumption of approximately 25% enrollment for the people in the university age bracket. The proportion of student category to the whole population in the 19-21 age group was 14% in 1969, is expected to go to 20% in 1975, and to 25% in 1978.

The problem faced by West German education policy is clearly visible. If the Universities do not expand, then somewhere between 20-40% of those willing to study will not be able to find a place in a West German university. This is contrary to the West German law which guarantees a university education to all applicants who have satisfactorily completed their prior training. In essence the solution will probably combine more stringent entrance requirements with some capacity increases.

The government plans on spending increased amounts for higher education and this is generally shared equally between the individual states and the Federal Government. The financing of post graduates by the government is presented in Exhibit 7-7. The figure for 1971 was 28.1 million dollars and is

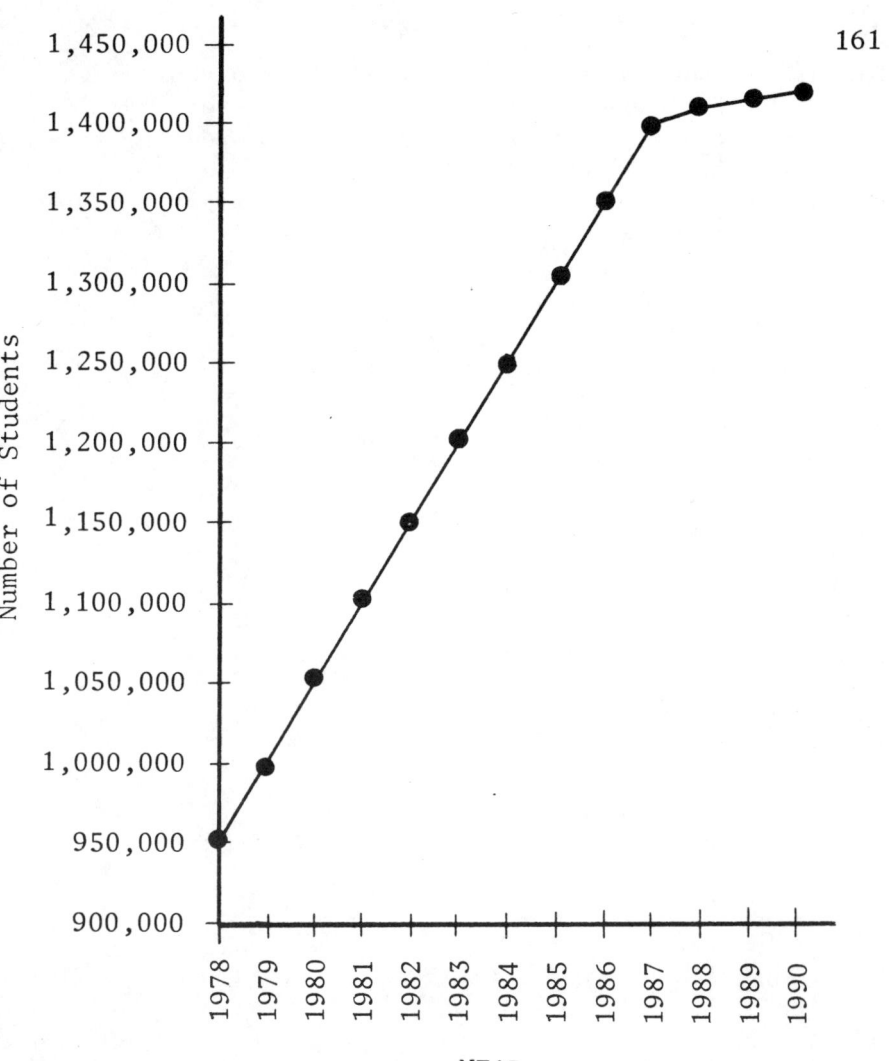

Number of Students in West German Universities*

Exhibit 7-6

*These figures are from a paper presented by Dr. Horst Albach, Bonn, "Models of University Planning as a Tool for Education Policy in the Federal Republic of Germany." The paper was presented at the European Institute for Advanced Studies in Management, Brussels, November 1974.

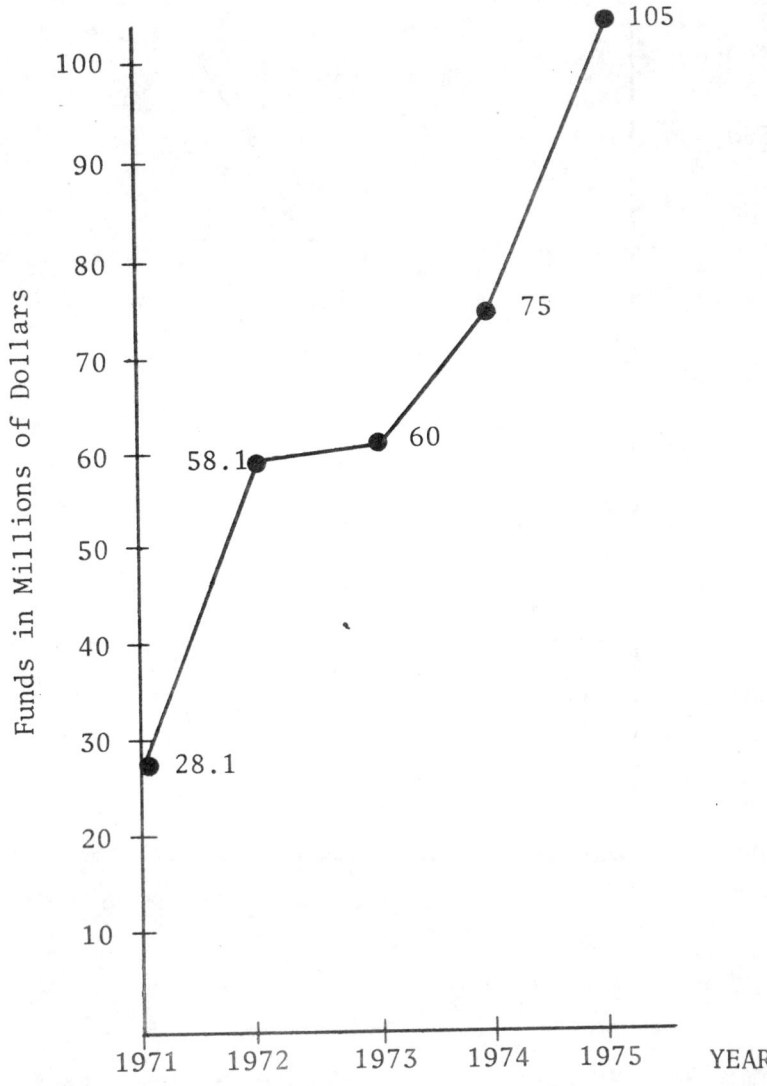

Government Financing For Post-Graduates (Doctorates, etc.) in West Germany from 1971 to 1975

Exhibit 7-7

expected to reach 105 million dollars in 1975. As one can see, the increase between 1971 and 1975 for post graduate support is approximately 250%. Even with today's inflation rate this clearly shows a strong commitment on the part of the Western German government.

The complaint about finding qualified engineers was heard quite frequently. Perhaps these complaints have acted as the stimulus for the increased educational effort. For many projects, large numbers of engineers and scientists are used and their supply has been limited. There is no longer the problem of "brain-drain" where engineers were leaving the country. The problem is merely one of supply.

At present there is little emphasis on the arts in technical training. The dilemma that the universities find themselves faced with is that of needing more specialization (as technologies become more complex) and at the same time needing more emphasis on social issues (see Chapter 6 on third generation engineers).

The current thinking is to emphasize the broad sciences and engineering and allow the specialization to take place within industry. The editor of a fine journal, Wissenschaftsstatistik (Science Statistics), pointed out that many engineers find themselves working in specialties that are different from their original training. In Chapter 6, we saw how the car is becoming increasingly electronic.

The universities themselves are in the midst of reappraising their role. The West German universities generally concede that in certain technical fields, the U.S. educational system is better than theirs, while in others it is poorer. Perhaps it was best summed up by one professor who has taught in the United States and in Germany who said that the difference between U.S. and German education in the sciences is that the best of U.S. cannot be beat, but the variance in U.S. education is very large.

The West German variance is small, and the West German standards are rigidly adhered to so that one school is very much like another. This is, of course, contrasted in the United States where the college awarding the degree largely determines the strength of the degree. In the United States there are very strong schools and very weak schools. In West Germany the applicant simply applies to a "Central Agency" and they tell him to which school he should go.

Having examined diffusion from the standpoint of the firm, the diffusion of basic research, and the universities role in diffusion, let us now try to put the pieces of the puzzle together, and examine the entire innovation system in perspective.

Chapter 7

Suggested Readings

Aiken, Michael, and Jerald Hage, "The Organic Organization and Innovation," Sociology, 5, 1971.

Allen, T. J., "Communication Networks in R & D Laboratories," R & D Management, Oct. 1970.

Allen, T. J. and Alan R. Fusfeld, "Design for Communication in the Research and Development Laboratory," Technology Review, Vol. 78, No. 6, May 1976.

Johnston, R., and M. Gibbons, "Characteristics of Information Usage in Technological Innovation," IEEE Transactions on Engineering Management, Vol. EM-22, Feb. 1975.

Mansfield, E., Technological Change, New York: W. W. Norton, 1971.

Paulson, S. K., "Causal Analysis of Interorganizational Relations: An Axiomatic Theory Revised," Administrative Science Quarterly, Vol. 19, No. 3, 1974.

Pavitt, K. and W. Walker, "Government Policies toward Industrial Innovation: A Review," Research Policy, Vol. 5, No. 1, Jan. 1976.

Rothwell, R., C. Freeman, A. Horlsey, V. T. P. Jervis, A. B. Robertson, and J. Townsend, "SAPPHO Updated - project SAPPHO Phase II.," Research Policy, 3, 1974.

Thompson, James D., "Technology, Policy, and Societal Development," Administrative Science Quarterly, March, 1974.

CHAPTER 8

The Infrastructure Necessary for Innovation

Perhaps the most important argument for a systems conception of organization is that the environment within which organizations exist is becoming increasingly unstable. With the rapid growth of technology, the expansion of economic markets, and rapid social and political change, come constant pressures for organizations to change, adapt, and grow to meet the challenges of the environment.*

The purpose of this chapter is to synthesize the preceding material and, where appropriate, add the findings of other studies with the objective of presenting a general perspective on innovation systems.

As in other countries, the factors relating to the West German innovation system are exceedingly complex and depend on many interrelationships. Exhibit 8-1 attempts to put these interrelationships into some perspective. The government affects innovation in industry through direct and indirect means.[1] Similarly, the government affects legal, financial, and educational institutions and their roles in the processes in innovation. As shown in Exhibit 8-1, industry affects the legal, financial, educational, and governmental decisions affecting technology. The exhibit shows this complex mix, and the absence of any ingredients will be a serious detriment to the

---

*Edgar H. Schein (M.I.T.), Organizational Psychology, 2nd ed., Prentice-Hall, Englewood Cliffs, N.J., 1972 (p. 104).

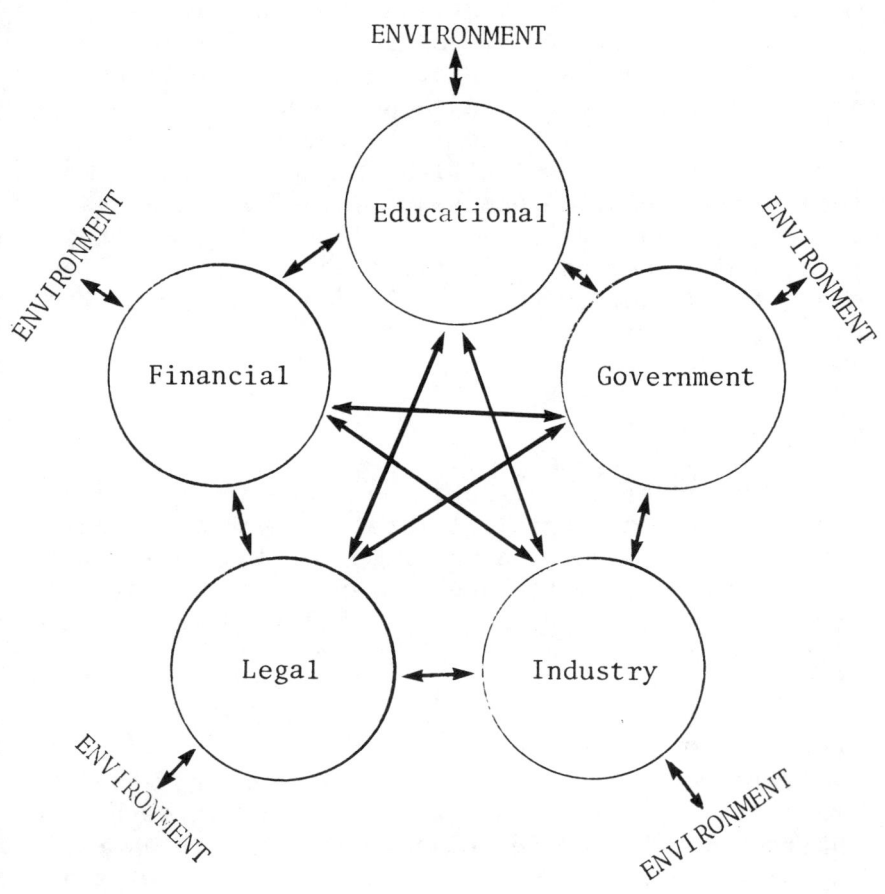

A General Model of the Organizations
Affecting Innovation: A Complex Interrelated System
Exhibit 8-1

innovative process as a whole.

We are beginning to understand that in West Germany and in other countries, mere government expenditures on R & D will have only a minimal effect on national innovation unless the other elements of industry, legal, financial, and educational are also involved. Similarly, the increased emphasis on industry must also be coupled with appropriate government incentives, legal mechanisms, financial resources, and educational support. To gain the proper perspective on West Germany's institutions and their affects upon innovation, let us briefly examine each of these elements.

Government

The government cannot legislate innovation any more than it can legislate morality. What it can do, however, is to become a major factor in the innovative process mix. The government plays a very important part in West German innovation through direct and indirect means.

The direct expenditures in research and development by a nation are necessary to stimulate innovation in fields such as national security, energy, health, etc. Although essential, direct expenditures have inherent inefficiencies since vast sums are awarded to industry regardless of the success or failure of a given innovation. Both the winners and the losers are rewarded.

West Germany is the largest spender of funds for research and development after the United States. The amount of funds spent by West Germany on research and development have been consistently increasing. The ratio between the government's contribution and industries' contribution in West Germany is approximately 50-50.

The major government expenditures for R & D in West Germany are for nuclear energy, defense, space, and electronic data processing. Other areas covered by government funding for R & D include health, basic science, communications, and non-nuclear energy.

In addition to the direct expenditures for R & D, the West German government affects innovation through indirect systems such as regulations and taxes. In Chapter 6 we observed how safety regulations are stimulating new dyes and filters.

Both the Northwestern and the M.I.T. studies show that projects resulting from governmental regulations are generally more successful than other projects.* The data indicates that environmental and safety regulations are often a stimulant to successful innovation. This is shown in Exhibit 8-2. Those projects which are authorized in direct response to government action have a smaller positive correlation with success.

Industry

The hub of the innovative process in West Germany and other Western countries is industry.[2] We know that in West Germany as well as in other countries that there are many more unsuccessful projects than there are successes. It behooves governmental and

---

*M.I.T. Five Country Study including West Germany, U.K., France, Netherlands, and Japan. N.S.F. Office of R&D Assessment, 1975; Northwestern. Albert H. Rubenstein, et al., Final Report on Field Studies of the Technological Innovation Process, Northwestern University, Evanston, Illinois, September 15, 1974.

Exhibit 8-2

Characteristics Related to Project Success vs. Failure in M.I.T. Study & Northwestern†

Government Factors	M.I.T. (U.K., France, Netherlands, West Germany, & Japan)	Northwestern (U.S. cases)
Environmental and safety regulatory constraints perceived to be highly significant	***	***
Project authorized in direct response to government action	*	*

*** A strong positive relationship ($p < .01$)

* A slight positive relationship but not statistically significant

---

†Northwestern reference: Final Report on Field Studies of The Technological Innovation Process, Rubenstein, Chakrabarti, and O'Keefe. Northwestern University, Evanston, ILL, September 15, 1975; M.I.T. reference: National Support for Science and Technology: An Evaluation of Foreign Experience, Center for Policy Alternatives, CPA 75-12, M.I.T. Cambridge, MA August 18, 1975.

industrial decision makers to develop early warning detection systems so that these potentially unsuccessful ones can be identified. We learned from the West German data that failed projects often run as long as successes. We must identify the predictive variables so that these projects can be identified sooner and technological effort directed more fruitfully. An effective early warning detection system is invaluable for the company and country.

The web of innovation becomes even more complex when we recognize that, in addition to industries' interrelationships with other elements such as government, finance, education, etc., there is now the recognition, based on the West German data, of the importance of industries' interrelationships with other industries. Steps must be taken to aid in industries' intercommunication. Innovation breeds innovation where a labyrinth of industries exist. But there are enormous difficulties in stimulating innovation where the labyrinth of industries does not exist. We observed in West Germany example after example how successful innovations involved the cooperative efforts of many firms. As described in Chapter 5, fifty percent of the research on the shock absorbent auto bumper was performed by a major plastics company, and glass companies' and electronic companies' research efforts aided in the development of thin film circuits. The chemical company developing a dye for non-shrink wool used textile companies for portions of the research.

We have observed from previous data (further supported by this study) that innovations that result from demand-pull are generally more successful than those from technology push. Similarly, Teubal found in Israel that the differential performance of R & D programs is explained by differences in the weight of the market factors versus technological factors. It argues strongly for porous organizational boundaries. The "radars" or antennaes of firms must be constantly tuned to the outside market.[3] As change becomes ever more rapid, it will become increasingly important to fine tune the sensing devices to aid in the recognition of necessary demands.*

We can expect to see both product and process innovation in the future. The spiraling labor costs will clearly encourage process innovations. Concomitantly, the brisk rate of change will influence product innovation. Influencing the product process decision is the role of competition and the incentive of high profits during early product life.

Exhibit 8-3 shows that when there is a specific competitive stimulus for a project there is a much higher probability for success than without such a stimulus. The data shown in Exhibit 8-3 is from West Germany plus four other European countries.

Legal Systems

The problem of employed inventors, as discussed in Chapter 2, is one that is not to be dismissed lightly. The West German model of the inventors law is an interesting example of a system developed to provide appropriate incentives for employed inventors. Clearly, as the years go on, we will witness more inventions from teams of individuals, and as costs rise, these teams will be company employees.

In West Germany the large companies particularly utilize the legal systems of patent protection. Chapter 3 showed how 53% of the actual discoveries would not have been embarked upon by the large corporations without the benefit of patent protection. This

---

*Gerstenfeld, A., "Techniques of Technological Forecasting," Journal of Business, University of Chicago, 1970. The necessity for firms to evaluate future technologies becomes increasingly important as the rate of change increases.

Exhibit 8-3

Competition Related to Project Success or Failure
(from M.I.T. five country study)*

Primary Stimulant for Innovation	Successful Projects (n=66)	Unsuccessful Projects (n-51)	Level of Significance of the Difference
Innovation as a result of specific competitive stimulus for the project	42.9%**	23.9%**	0.07

*M.I.T. Reference: National Support for Science and Technology: An Evaluation of Foreign Experience, Center for Policy Alternatives, CPA 75-12, M.I.T. Cambridge, MA, August 18, 1975. The five countries included West Germany, U.K., Netherlands, France, and Japan; NSF R&D Assessment Program 1975.

**The percentage means that 42.9% of the 66 successful projects studied, the project managers' stated that the particular innovation was started as a direct result of specific competitive stimuli. This can be contrasted to the unsuccessful projects where only 23.9% of the 51 projects were started because of specific competitive stimuli.

was not true for the smaller companies where only 3% of the actual discoveries would not have been realized. Perhaps this is due to the high cost of legal services in the West German innovation system or perhaps it can better be explained by the investment size. There is no doubt that the R & D efforts of the larger firms in West Germany involve larger expenditures; hence, encouraging the benefits from patent protection.

The patents studied in West Germany show that the smaller companies convert their discoveries to economic use in a much shorter time than the larger companies. The legal protection of the patent enables the larger companies to often take up to four years before converting their discovery to economic use. Of course, one could argue that these patents are more complex and therefore take longer to reach the market, but nevertheless, the fact remains that the legal protection of patents in West Germany is important for both large and small firms, but more important for the larger firms.

Financial

The financial systems in West Germany are of course necessary for innovation and vary for each situation. We have already discussed the government financing of direct R & D as one method of stimulating innovation. Government funding is also needed to stimulate innovation in areas of social concern or national welfare. At present, in both the U.S. and West Germany, industries' internal funding for R & D approximates the governments outlays. Most non-military innovation is supported in this manner.

In Chapter 4, we examined the funding for small and medium sized business innovations in West Germany. In a recent analysis on corporate size, it was pointed out that 5 out of the largest 15 European corporations are West German firms. The companies mainly finance their own R & D.

In medium and small sized companies in West Germany, the innovation is also often financed by internal funds. However, it is also recognized that medium and small sized companies have difficulty in financing innovation. As a result, the Reconstruction Loan Corporation often finances innovation for these sized companies with particular emphasis on new processes.

The new risk financing and venture capital organization forms the financial base for innovations from new enterprises in West Germany. The banks contribute a certain amount and the Ministry of Research and Technology guarantee 75% of all losses.

We therefore see several vehicles for the financial community to play a role in the innovation system of West Germany.

Education

The role of higher education in the innovative system is two-fold. The universities have the responsibility to perform basic research, and concomitantly, the responsibility to supply skilled technical manpower.

Basic research is generally performed by the university through an interrelationship with the government. This basic research affects innovation in one of two ways. It often forms an underpinning for the more applied research and development. Conversely, R & D often leads to basic research as in the West German nuclear energy programs which stimulate investigations on sources of matter. Only the few largest corporations can afford to do basic research.

West Germany recognizes that it is equally important that it maintain the supply of skilled scientists such as physicists, chemists, biologists, etc. Similarly, the supply of engineers and technicians must be adequate to meet the coming demands.

From the data presented in Chapter 7, it can be seen that the universities in West Germany are seldom used as a direct information channel affecting innovation. The linkage remains less direct and more as a manpower source and performer of basic research. The universities do participate with the industrial research associations on some joint innovation processes.

Environment and Diffusion

The environment as shown in Exhibit 8-1 completely surrounds the innovative system. It is the environment that affects governmental decisions, and similarly, governmental decisions affect the environment. As shown in Chapter 6, social values change and rapidly affect portions of the innovative system.* The environment includes not only the physical but the market as well. The innovative system responds to the market and the forces of the environment which encompass it.

Throughout all the portions of the innovative systems, I have stressed the importance of the interrelationships as shown by the arrows in Figure 8-1.

Perhaps these interrelationships can best be visualized by considering the parts of a

---

*Utterback's work has emphasized the importance of the external environment on the organizations' innovation process. Utterback, James M., "The Process of Technological Innovation Within the Firm," Academy of Management Journal, 14, No. 1, 75-88.

fine watch all placed into a cup. There can be jeweled bearings, fine springs, an engraved face, and well-made minute hand and hour hand. Nevertheless, unless these parts are placed together in a meaningful way, we do not have a working system. This analogy applies equally well to the technological system. The connections among education, government, financial, legal, and industry must be accurate and reliable. The technological system, like the fine watch, must have all good functioning parts, but, equally important is the linkage among these parts.

Without diffusion, the entire system comes to a halt. The diffusion of information in the basic sciences is good. Meetings are held, papers are published, and generally the flow of information is quite effective.

This is not true for the applied areas, and the flow of information remains as a major problem in the innovative system. Companies in West Germany and elsewhere still have difficulty communicating within their own organizations, and one project group often has little knowledge of information available from another project group.[4] The problem is multiplied between companies, and exacerbated between industries. In the coming years, increased emphasis on this portion of the innovative process could reap substantial benefits.

We have examined many aspects of innovation, with the goal of starting to gain a better understanding of how innovation takes place. The process is complex, parts are interrelated, and it is dependent on many variables. It is hoped that these concepts start to identify some of those variables and the way in which one relates to the other.

Part II

Case Studies in Innovation, Research & Development, and Technological Policy

For those readers using this book in the classroom, it is suggested that the following cases be used for analysis and discussion in order to illustrate issues raised in the various chapters. Each case relates to material from several of the chapters, however the closest relevant chapter to the topic will be listed although the case analysis need not be limited to that subject. The first and third cases use actual company names since the material has been released publicly in a different form. The second case is based on an actual situation but uses disguised names to maintain anonymity.

Case Name	Major Topic	Chapter
1. The $100 Million Object Lesson	Factors related to successful innovations: demand pull, technology push, level of effort, and early warning systems	5
2. International Production Technology, Inc.	Funding for Technology: the role of U.S. spin off corporations in high technology	4
3. Innovation at Texas Instruments	Social Forces Effect Upon Innovation: planned innovation, innovation and corporate growth	7

Case Study 1

THE $100-MILLION OBJECT LESSON*

The road from laboratory to market place is strewn with pitfalls that occasionally trap even the most practiced user of research. Consider, for example, Du Pont's long, costly misadventure with Corfam - the synthetic competitor of leather that was once touted as "another nylon." After seven years of effort and an investment of close to $100 million, Du Pont has yet to earn its first dime from Corfam. Company officials openly concede that the product has been a severe disappointment. One high-level executive suggests that it has only another year or so to prove itself.

What went wrong? For all the agonizing reappraisals that have taken place within the company, nobody has a firm answer. "To say it was a marketing failure just isn't enough," says a member of Du Pont's executive committee and the vice president responsible for research liaison. "We really overestimated its value-in-use." An even more sweeping assessment comes from the original head of the Corfam program, who has since been promoted to assistant general manager of the fabrics and finishes department. He says: "Usually in a new project you encounter compensating errors. With Corfam all of the errors were on the wrong side, and they were important."

Du Pont's first fundamental research on porous polymeric film took place in its central research laboratory in the 1930's. Nothing was done to exploit

---

* From: Harvard Business School 3-672-016; Reprinted from "Bringing the Laboratory Down to Earth," by Dan Cordtz, c1971 by Fortune. Reprinted by permission.

these early discoveries, however, until the company's fabrics and finishes department began in 1950 to look seriously at the market for shoe uppers. After detailed marketing studies, the department concluded that by 1982 there would be a substantial shortage of suitable leather and that 30 percent of shoes would have to be made from an alternative material. Since the most popular leather substitute at the time, vinyl-coated fabric, had the serious flaw of impermeability (inability to "breathe" and dissipate moisture), the opportunity for profit seemed clear if a better product could be developed.

A Hint of Trouble

By 1962 the department's scientists, working closely with the shoe industry, had a material that met all of the perceived requirements. Some 15,000 pairs of shoes were made of the new material and tested under typical conditions by representative users. Their reaction was encouraging, especially their enthusiasm for the ease of caring for Corfam shoes. About 8 percent of the users complained of some discomfort, but so did 3 percent of those wearing leather and 24 percent of those wearing vinyl-coated fabric shoes. "Not too bad," an executive says. True enough - but the complaints, while not given much weight at the time, proved to be a hint of trouble to come.

Reassured by the research group's estimates of what large-scale production costs would be, Du Pont was confident of Corfam's ability to compete economically. Leather for uppers was selling in a wide range - from 30 cents to $5 a square foot. Corfam's lowest price was $1.05 a square foot, but its uniform quality and the convenience of its sheet form clearly merited some price differential over unwieldy and frequently uneven cowhides. The project planners believed that Corfam could take a healthy share of the market from leather in the price range of 80 cents to $1.05 a square foot. This restricted initial use of the material to shoes priced at $20 and up, but Du Pont hoped that Corfam's price might eventually be whittled enough to enlarge the potential market.

In the fall of 1963 some thirty shoe manufacturers showed Corfam models at the National Shoe Fair. Demand soared, and in August, 1964, Du Pont opened a new plant to make Corfam at Old Hickory, Tennessee. But technical problems had already begun to crop up and, with large-scale production, they grew rapidly worse. Making Corfam was and remains a complicated, highly sophisticated process. This is the main reason why Du Pont never showed much fear of competitors' copying the product, although Goodrich, Tenneco, and many smaller companies tried. In scaling up from pilot operations, the technical experts were never able to turn out Corfam at the predicted cost while retaining its essential competitive qualities. Since the market had to be allowed to set the price, red ink flowed out of the Old Hickory plant about as fast as Corfam.

"Aesthetic" problems also surfaced. Shoemakers and purchasers alike were dissatisfied with what in a fabric would be called "hand" - the feel of the material and the way it bends and creases as it flexes in use. Wearers raised the comfort question more persistently than earlier surveys had suggested they would. Du Pont technicians insisted that Confam was even more porous than leather, but many wearers complained that shoes made of the synthetic felt hot. And because the material was more resilient than leather and did not acquire a permanent stretch, Corfam shoes remained tight after prolonged wearing. With impeccable logic, Du Pont officials pointed out that this would be no problem if customers merely moved up to a slightly larger size. But this rational observation bumped into the irrational psychological fact that not many men, and fewer women, like to acknowledge that they have big feet.

Maximizing Virtue

Although Du Pont had some acquaintance with the shoe industry, from supplying Neoprene soles, nylon thread, cements, and other materials, it seems clear now that the company's marketing specialists did not appreciate exactly what they were getting into. If they did understand fully just how formidable a competitor leather would be, they apparently did not adequately communicate the problems to the research staff. And the scientists - justifiably proud of a remarkable technical achievement - evidently overestimated their product's virtues and minimized its faults. They also erred in their forecasts of how cheaply it could be manufactured. Finally, management, dazzled by the glittering commercial prospects, was not skeptical enough in the beginning and was slow to recognize the extent to which plans had gone awry.

Author's note:

[In June, 1976 I talked to Du Pont officials to determine the latest status of Corfam. I was informed that Du Pont has licensed the Polish government to manufacture and sell Corfam anywhere in the world with the exception of Japan and North America. A U.S. company is buying the semifinished material from Poland and finishing the goods for sale in the U.S. On the other side of the world, the Japanese have a license from Du Pont and will manufacture and sell in the Far East. Du Pont will be receiving royalties and attempt to recoup from the $100-million object lesson.]

Case Study 2

INTERNATIONAL PRODUCTION TECHNOLOGY, INC.*

Mr. Thomas "Stoney" Edwards, President of International Production Technology, Inc. (IPT), was concerned about the future of his company. Formed as a spin-off of the Equipment Division of Siliconix Incorporated (a semiconductor manufacturer) in June 1969, IPT had survived the effects of the economic downturn of 1969 and 1970 by supplying high quality advanced state-of-the-art semiconductor process equipment which provided high margin returns. The problem confronting Stoney was how IPT was going to maintain their high technology edge on a very competitive industry. Many companies like IPT had not survived beyond their frist generation of equipment.

Formation of IPT

IPT was a spin-off of Siliconix Incorporated (SI). SI was formed in 1962 to produce semiconductors, particularly field-effect-transistors (FET). Since SI was producing hardware on the frontiers of the state-of-the-art, it was necessary for them to design and build their own process equipment. This required the formation of an in-house equipment development group. The first major piece of equipment of this group was a wafer probe. The wafer probe, in addition to being a solution to one of SI's process problems, also turned out to be a highly profitable product. A number of other pieces of equipment developed by this group also turned out to be highly profitable products. This resulted in the equipment group being formed into SI's first division - the Equipment Division - in early 1967.

---

* From: Harvard Business School 9-672-066; Copyright 1971 (Rev. 6/72) by the President and Fellows of Harvard College. Reprinted by permission. Actual case with names disguised to maintain anonymity.

The Division was headed by Stoney Edwards who shared equipment development responsibility with Mr. Dan Worsham. Although the Division's management was charged with manufacturing and marketing responsibility for equipment products, product policy decisions were still in the hands of Dr. Richard Lee, the President of SI. Because of the primary concern of SI's management with component products, the management of the Equipment Division felt that decisions about equipment development were often made which were not in the best interest of the Division. Such a decision was reached in the winter of 1967 when Dr. Lee decided to have the Division build lead bonders, a piece of equipment needed by SI's semiconductor manufacturing process, rather than continuing the development of an automatic probe, a product which the Division's management thought would be highly profitable.

As a result of this decision, Mr. Worsham left Siliconix in early 1968 to form his own company in order to develop and market automatic probes. Mr. Edwards described how this conflict of priorities affected the rest of the Division.

> For some time the Division's priorities had been shifting to outside customers. It was this shift which pressured for the development of IPT. Dan's departure emphasized the need to recognize the changing priorities.
>
> I can't say whether I would have left Siliconix or not. I liked the company; I still do. I was, however, reaching a point where I felt we ought to commit ourselves fully or get out. This nibbling around the edges of the equipment business with no financial support or formal organization was pointless.

By mid-1968 Mr. Edwards and Dr. Lee had reached an informal agreement that if the right conditions and people could be brought together to form a separate company out of the Equipment Division, the board of directors would consider a proposal. The right people came together quite unexpectedly. In the process of looking for a replacement for his engineering manager, Mr. Edwards talked to Mr. Chuck Bodine - chief mechanical engineer for Signetics Corporation. The two men found that they had similar objectives in starting a new company. Mr. Bodine and a Mr. John Day (who was with Itek Corporation) had already bid on a product which Itek was trying to sell to improve the company's cash flow, but the two men had been outbid by Texas Instruments. Mr. Edwards described how the three men became the founders of IPT.

>Chuck and John were in a state of suspension; they wanted to put something together but didn't have anything at hand. Chuck liked the idea of starting with a base organization like our Equipment Division. We both worked on John to sell him on the idea of using Siliconix's Equipment Division as a base. Once we agreed we approached Dick Lee with the idea.

The three men worked together at nights from September 1968 to April 1969 putting the concept of IPT together. In October 1968 Stoney introduced Chuck and John to Dick Lee and the three men made their first formal proposal in December. An agreement in principle was reached in February 1969 and in March the three men started looking for financing. On April 1, 1969, Messrs. Bodine and Day quit their jobs and became consultants to SI (after IPT was formed the new company reimbursed SI for the two men's consultant fees). A formal agreement was reached in May between the founders, SI and investors and IPT was incorporated under California law on June 19, 1969. In exchange for 40% of the stock (in convertible preferred - a form of tax-free reorganization) of IPT, SI transferred to the new company physical assets valued at $364,000 and all equipment designs, customers and sales. In addition, SI maintained two members of IPT's seven-man board of directors. In exchange for $1 million the investors acquired about 30% of the company and the founders and key employees purchased 30% ownership which could be repurchased by the company upon premature termination of employment.

The people in the new company were of two types - those who had come from SI and understood the business, and those who had come from outside of SI with new ideas they wanted to exercise but who had to learn the business. Stoney Edwards grew up with the equipment business since SI's formation in 1962. John Day had a MS in Mechanical Engineering from MIT and a M.B.A. from Stanford and had been the marketing manager for Itek Digital Systems Division. Chuck Bodine was a graduate mechanical engineer and had been chief mechanical engineer at Signetics. Prior to that Chuck had been a design engineer with Fairchild and Westinghouse. Not all of the people in SI's Equipment Division had joined IPT; some had not wanted to leave SI and some were asked not to leave. As a result of this mixture of people in IPT and SI, Mr. Edwards had some difficulty in maintaining good relations between the two groups.

As a result of the new mixture of people, there were strained relations between the operating personnel of IPT and SI during the first year. I had to wet-nurse the relationships for a while. I think our new people understand the needs of Siliconix and the Siliconix people understand that IPT is an independent business and must make decisions based upon the needs of IPT. If a Siliconix request doesn't make sense for IPT, we'll tell SI to go elsewhere.

IPT inherited four products from Siliconix - the wafer probe, FET tested, die bonder, and contact printer. The sales volume of the last year of the Equipment Division was $800,000 up $200,000 from the previous year. The sales volume for the first year

of IPT was $1.4 million, with a customer list of over
two hundred and with 30% of the company's business
overseas.

To maintain the growth rate of IPT, management
had developed a three-phase product strategy. Phase
one was to establish a product organization and
solidify the existing product line. This phase would
take six to nine months. Phase two would last from
one to two years and somewhat overlap phase one. In
phase two, IPT would extend the existing technology
within the existing product lines or similar products,
bringing in new products to do old jobs better or
supplement the existing products. Phase three would
deal with new product line development which would
result in IPT entering new markets which have different selling patterns. In order to implement this
product strategy, Mr. Edwards believed it would be
necessary to develop a technically strong well-
structured electrical-mechanical business.

> One of the things I'm trying to build into IPT
> is a higher level of scientific content than most
> companies in this field possess. Most equipment
> design is by draftsmen who came up through the
> ranks and they design everything by cut and
> try. I'd like to get some engineering talent to
> apply their knowledge of theory and design
> experience to the development of optimal equipment
> design.

Relationship Between IPT and Siliconix

Both Dr. Lee and Mr. Edwards described the
relationship between IPT and Siliconix as one of
supplier and customer. IPT had a salesman who called
on Siliconix but IPT did not provide all of SI's
equipment. In addition to equipment sales, IPT
had a machine shop contract with Siliconix under which
IPT supplied a stated number of man-hours of machine
shop time to SI. Relations were close enough that
SI would demonstrate IPT's equipment in their line to

potential customers. In addition, IPT would pick up the development of certain equipment on which SI engineers had developed to the breadboard design stage. Mr. Edwards described how this latter arrangement worked.

There have been a couple of areas where both companies have done some overlapping development work and Siliconix needed some of the equipment built. We wound up with a special arrangement whereby we built the equipment for them at cost in exchange for their part of the design. We put some of our knowledge together with some of their's and came out with a new product that we could sell to the market. We are essentially providing a service to them by designing and building an advanced product.

Problems Facing IPT

Dr. Lee and Mr. Edwards described some of the problems they saw facing IPT.

Mr. Edwards: Top management of a technically oriented company must have an understanding of the effects of technology. All decisions are not related to dollars and cents decisions. Profitability in any one decision must be looked at as part of a longer range objective. The primary operational problem is to keep creative individuals creating. The further out one looks the less important the sales department's point of view becomes and the more important the views of development engineers become. My role is to keep a balanced point of view. My problem as an engineer is to insure I am not overstressing my point of view.

Dr. Lee: Most equipment companies in the semiconductor industries are spin-offs from semiconductor companies of a couple of engineering oriented guys who were successful in designing in-house equipment for one of the giant semiconductor companies. They didn't feel there was enough return from the company and decided they could get rich quicker by starting their own company. The first thing they do is to duplicate a piece of equipment they designed in-house. From the success of that equipment they develop, from scratch, their next model of equipment. This is one they think the semiconductor industry needs. Meanwhile they are two years detached from their parent company and therefore two years detached from the industry's real needs. This becomes the source of the decline of this equipment company.

Mr. Edwards: One of the problems an independent equipment company faces is that he can't take the black box he's developed out on to his own line to run and debug like an in-house group can. The independent has to ship it out of house and there's an irate customer who isn't going to pay his bill until his equipment is fixed. It's a $10,000 piece of equipment and it's important to the little company to get it fixed both to get the revenue and to keep the customer's good will. This is a serious problem in the evolution of equipment companies.

Dr. Lee: The IPT spin-off was necessary to attract the right calibre of management necessary for the growth of the equipment business. We didn't want to sit on the business and choke it off, so we decided to keep a piece of the action but let it stand on its own two feet.

Mr. Edwards: In our operations we have not been influenced by Siliconix, outside investors or anyone else. If for any reason we do not survive, it will be our own fault.

Case Study 3

INNOVATION AT TEXAS INSTRUMENTS*

Since its inception in 1930 innovation has been a way of life at TI. In 1969 TI invested $160 million in research and development of new products, process development, mechanization design, and other technical support of manufacturing and marketing. S. T. Harris describes the reason for TI's dedication to innovation.

> We are convinced that useful products and services as well as long-term profitability are the result of innovation. Further, we feel that profitability above the bare compensation for use of assets can come only from a superior rate of innovation and can no longer exist when innovation is routine.
>
> This is why our long-range planning system is fundamentally a system for managing innovation.

The long-range planning system to which Mr. Harris was referring was the Objectives, Strategy and Tactics or OST system. According to Mr. Harris the OST system is a statement of existing practices which developed at TI.

> Almost ten years ago we identified a pattern that eventually evolved as the Objectives, Strategies and Tactics system by which we now manage innovation. The germinal idea came in recogni-

---

*From: Harvard Business School 9-672-036; Copyright 1971 by the President and Fellows of Harvard College. Reprinted by Permission. The major source of information for this case was a series of talks by Messrs. Harris and Doyle presented at the London Graduate School of Business.

tion of the fact that our previous successes as a company were based on well-conceived strategies. By strategies we mean those general courses of action the executive responsible intends his organization to pursue in achieving company goals. And supporting those strategies we identified specific programs that we now call Tactics, which had to be carried out to implement the strategies successfully.

For example, TI's great success in semiconductors was the result of a strategy accomplished through successful tactical R&D programs. Our strategy in the early 1950's was to seize the lead in the semiconductor industry from other much larger companies which might more logically have been expected to be out in front.

Our successful tactics, although we did not use this specific nomenclature for them, were (1) the establishment in 1953 of our materials-oriented central research laboratories; (2) the introduction of the first silicon transistor in spring of 1954; (3) the work which led to the first all-transistor pocket radio in October 1954, and (4) the development of a process for making ultra-pure silicon and the beginning of commercial production of that material in 1956.

A year later we initiated the OST system to identify and state succinctly, yet completely, in writing, the strategies we would follow for growth and development throughout the corporation. We also identified the tactics we would pursue to implement the strategies. There is no question but that we have succeeded in diffusing throughout our management at every level a recognition of responsibility for initiating innovating programs as well as an improved ability to conceive, describe and pursue such programs.

Innovation and Corporate Growth

One of the problems recognized by TI management was the difficulty which large organizations have in maintaining a high degree of innovation as they grow. It often happens that a given amount of innovation requires increasing amounts of corporate effort as the company grows. The balance of this case is an excerpt from Mr. Harris' speech in which he discusses his feelings about this problem.

As the organization grows, it has more resources, more knowledge, more qualified people, more contacts with customers, more opportunities and greater need for innovation. It should be getting easier, not harder. Why not? What's happening?

I can only tell you from my own experience what I think happens. The validity of my judgment you must confirm or deny from your own observations. At any rate, here are what I believe are some of the reasons:

As the organization grows, it gets more complex. Hundreds and then thousands of people are involved, often in multiple locations. The number of customers grows. Operations extend into many states and often into many countries. Governments all over add complexities of reporting and of regulation - some of them necessary, some of them not. To exploit an invention or innovation fully and to get broad distribution, the price must come down. The margin between price and costs gets narrower. At a relatively early stage in the development, so far as this invention or innovation is concerned, it becomes far more important that the principal managers be good administrators than good innovators. The administration in a technologically based business may often require good, or even deep technical skill, but at this stage what counts is the aid that kind of knowledge brings to administration, not to innovation.

For the reasons mentioned, most of the status symbols, such as department managerships, administrative staff managerships, officerships, and prerequisites, go to the administrative managers. The innovators around the organization begin to believe, with some justification, that the way to get promoted is to succeed at administration. Some do succeed and become good administrators. Down the road, their knowledge of both innovation and administration may make the crucial difference in whether the organization stagnates or continues to innovate and grow. However, because of the system many of these men won't fit the pattern that the administrative manager recognizes. Too rarely will they get the time and the right kind of experience to succeed at administration. Too often will they top out well down in the management hierarchy. Or if they do progress, they have to work frantically to stay even just because they are not good at administration. They are left with little energy or opportunity to innovate.

To handle the growth and increasing complexity, the organization decentralizes into groups, divisions, departments, and branches. The total job is divided up and cut into the size pieces that a good administrative manager can get his arms around. This is logical and good management practice. But unless the general managers understand their jobs thoroughly, the company is in danger of becoming no more than the sum total of the decentralized parts, loosely governed at the corporate level, primarily from a financial point of view.

At that point, the biggest job the corporation can handle has to be related to the biggest job that one, or but a few of the decentralized working units can handle working together.

The only way all this can be prevented is to tie the decentralized entities together strongly at the top. Topnotch general managers, aided by strong functional organizations in marketing, research and development, personnel, control, etc., must knit the decentralized line units tightly together. Every manager must learn to disbelieve the favorite management dictum that responsibility and authority always go together.

The right rule is that responsibility and authority must always go together to the maximum extent possible.

In a decentralized organization the span of responsibility almost always exceeds the span of authority. Each manager has an authority which extends only to his own decentralized units, but a responsibility which extends across the corporation. Even when good innovative managers do develop in a decentralized organization, their innovations usually are restricted to the entity for which they have responsibility, or only narrowly beyond it.

Consequently, though the organization as a whole may have far more of the tools, the opportunity, and the skilled people needed for innovation, the exposure of any one manager is restricted. He simply fails to see the larger opportunities to solve problems of the right scale for the whole corporation.

Another undesirable pattern which too often develops is the gravitation of resources toward short-term problems with a consequent neglect of long-term, major impact programs. As a result, the organization can be misled into believing that it is making sound investments for the future, when, in truth, the resources may be largely consumed in responding to current crises.

These are formidable challenges, and the remedies are not simple. TI does not have all the answers, nor even enough of them. But what we have done is to provide explicitly a mechanism for both operating and strategic modes in our organization.

Our operations are organized into a number of

relatively small businesses, which we call Product-Customer Centers. Each of these is operated by an entrepreneur with profit and loss responsibility for the create, make, and market functions. Thus, the manager has a clear view of the causes and effects over which he has near-term control.

At the same time, a strategic role is superimposed over his operating responsibilities. Normally, but not always, the strategy manager and the operating manager are the same individual. When they are, it simply means the manager is expected to wear two hats, and is given the proper goal and policy statements to balance the two roles. When the strategy and operating managers are not the same individual, this simply means that a conflict situation has been intentionally structured to force key issues and decisions higher in organization. Obviously, this particular mechanism must be used sparingly because of the demand it makes on top management attention.

Even so, our approach does require an intense level of top management commitment. I would certainly caution you about this commitment - not just in dedication to the organization and its goals, but in a willingness to devote the necessary time to clarifying objectives, how you propose to reach them, and in what areas initiative will be acceptable.

One thing is certain: few things can paralyze an organization's effectiveness as uncertainty in the top management's goals and policies. Our Objectives, Strategies and Tactics system is designed to remove these uncertainties and replace them with a well-defined and well-communicated goal structure extending well down into the organization.

NOTES

CHAPTER I

1. The United States Senate recently passed the "National Energy Research and Development Act of 1973" (S-1283). S-1283 would establish a national R & D program in fuels and energy in which the Federal Government would provide about $20 billion during the next 10 years.

2. The Second Report from the U.S. Department of Commerce/Patent Office shows these data (Early Warning Report of the Office of Technology Assessment and Forecast, December, 1973). Their data show about 30% of U.S. patents being issued to residents of foreign countries of which West Germany is the highest.

3. Schumpeter visualized technological change as occurring in three steps: invention, innovation, diffusion (or imitation). Joseph A. Schumpeter, "The Theory of Economic Development" (Cambridge: Harvard Univ. Press, 1934). The trouble is that it leaves ambiguous the lengthly research and development sequence of detail technical activity.

4. Helmar Krupp, (Director, Institute for Systems Engineering and Innovation Research (151), Karlsruhe West Germany) made this point particularly during a talk in Brussels on November 7, 1974.

5. The Report on Research (IV) of FRG, 1972 points out that industrial organizations account for half of R&D expenditure in West Germany. The actual figures are from a paper "The R&D System in FRG and Innovation Research to Improve Its Performance." The paper was authored and delivered by Dr. Kramer, Ministry for REsearch & Technology, Bonn, Germany. The paper was delivered at George Washington University on September 25, 1974,

sponsored by NSF. The 1974 figures are estimates but believed to be quite close to reality.

6. OECD, Paris, May 1974, "Patterns of Resources Devoted to Research and Experimental Development, pg. (V).

7. OECD, Paris, May 1974, "Patterns of Resources Devoted to Research and Experimental Development, pg. 10. The facing page explains, "R&D activities grow more rapidly than total national resources in all the countries except the United States and the United Kingdom."

8. The percentages of government support for U.S. are from the Statistical Abstract of the U.S., Dept. of Commerce, 1973. The percentages of government support for West Germany are shown in the paper by Kramer, Ministry for REsearch & Technology, Bonn, Germany, delivered at George Washington University, Sept. 1974.

9. U.S. Data from "An Analysis of Federal R&D Funding by Function," NSF 73-316 (Oct. 73). West German data from Figure 4 of previously cited Kramer paper. The percentages for '72 show defense as 12.7% of total R&D. For estimated '74 figures the percentage was kept the same and the absolute amount increased from $0.5 billion to $0.6 billion.

10. There is often a long lag time between the military product and the civilian application. The use of sonar for detecting birth defects is finding importance in 1976, 25 years after its use as a submarine detection device.

NOTES

CHAPTER 2

1. The Institute for Economic Research in Germany recently investigated the general patent law, and the inventors law. The investigation showed that actually 94% of all patents of German inventors are employees' inventions.

2. Simon, Herbert A. "Theories of Decision-Making in Economics and Behavioral Science, "American Economic Review, 1959; Simon refers to "psychic income" from the firm, quite apart from monetary rewards. In this case he's referring to entrepreneurs, but it's clearly applicable, if not more so, to the employed inventors.

3. The Statistical Abstract, U.S. Department of Commerce, 1973 shows a decrease in patents applied for, 1971 - 111,095 and 1972 - 105,300. For patents issued a similar decrease is shown as follows, 1971 - 81,789, and 1972 as 78,183.

4. A survey of the laws on employees' inventions was the discussion of a conference in Feb. '71 in Zurich, see F. Neumeyer, "The Employed Inventor as Subject of Legislation - an Ideological Survey," Industrial Property, 1971.

5. Scherer, F.M. "Industrial Market Structure and Economic Performance," Rand McNally, 1970 - Scherer describes governments as granting exclusive patent rights for three main reasons: to promote invention, to encourage the development and commercial utilization of inventions, and to encourage inventors to disclose their inventions to the public.

6. The Journal of The Association for the Advancement of Invention and Innovation has pushed hard to reestablish recognition of the importance of U.S. invention and innovation. See Action, AAII, Arlington, Virginia, Jan.-Feb. 1974.

7. Dr. of Law, Hans Schade. From 1957 - 1971, he was Chairman of the Arbitration Board set up within the German Patent Office by the Law on Employees' Inventions. The article is entitled, "Employees' Inventions - Law and Practice in the Federal Republic of Germany," Industrial Property - Sept. 1972.

8. T.J. Allen, A. Gerstenfeld, and P. Gerstberger, "The Problem of Internal Consulting in Research and Development Organizations," M.I.T. Working Paper 319-68, July '68; In this paper we show how cross-fertilization of ideas and the mixture of specialization in various areas of expertise generate innovation and creativity.

NOTES

CHAPTER 3

1. Vernon, Raymond, "Manager in the International Economy," Prentice-Hall, N.J., 1972. Vernon points out the following: "How to secure the benefits that are sought through the grant of patent monopolies without their potential drawbacks is one of the unresolved problems of governments that the manager will encounter repeatedly in the international environment."

2. Richard M. Cyert and James G. March, "A Behavioral Theory of the Firm (Englewood Cliffs; Prentice-Hall, 1963), pp. 278-79; Cyert and March contend that innovation is the result of a conscious search for new and better solutions to pressing problems.

3. The concepts of patent monopoly is discussed in an article that was part of an N.S.F. study of the patent-antitrust interface. John C. Stedman, "Patents and Antitrust - The Impact of Varying Legal Doctrines," Utah Law Review, Volume 1973, Winter, No. 4, p. 589.

NOTES

CHAPTER 4

1. The importance of a well-working capital market is stressed by several authors. Prodi, R. The Diffusion of Innovation in Italy, Bologna, 1971.

2. Some innovations have become very expensive and require a concentration of resources. See Layton, C. "European advanced technology: a program for Integration." P.E.P. - Allen & Unwin, 1969.

3. John Kenneth Galbraith, "The New Industrial State," Houghton-Mifflin, Boston 1967 (p. 4). Galbraith cites the increased time required for development of todays technologists as follows: "They (sophisticated technology) involve a greatly increased elapse of time between any decision to produce and the emergence of a salable product."

4. Jesse W. Markham, "Market Structure, Business Conduct, and Innovation," American Economic Review, May 1965, p. 325, Markham finds that there is a threshold effect. He claims that roughly $75 to $200 million in most industries is good for invention and innovation. Beyond that threshold, further bigness adds little and carries the danger of diminishing the effectiveness of inventive and innovative performance.
(I feel that we've seen exceptions on both sides).

5. Scherer, F.M. "Industrial Market Structure and Economic Performance," Rand McNally, Chicago, 1970, p. 350. Scherer describes the investment in technological activity as the act of risking funds for the venture.

6. Scherer, F.M. "Industrial Market Structure and Economic Performance," Rand McNally, Chicago, 1970, p. 350. Scherer points out that the functions of technological activity and the financial support need not be performed by the same organizational entity; and in fact are often organizationally separate.

7. The association between the financial community and innovation was investigated by Cacace. Cacace, N. "Innovation of Italian Products," Milan, 1970.

NOTES

CHAPTERS 5 and 6

1. For this portion of the research I particularly acknowledge Prof. J. Allen (M.I.T.). The M.I.T. Center for Policy Alternatives stressed the "specific project approach" which enables the researcher to get beyond opinion and into some data analysis.

2. a) Statistics on Science and Technology, 1970, Dept. of Education and Science and Ministry of Technology, p. 116;

   b) Doina Thomas: 'Paint Makers' Problems, Marketing, Nov. 1970, p. 43.

   c) Philip Kotler: Marketing Management, Prentice-Hall, 1967, p. 79.

   d) Sir Charles Goodeve: "A Route 128' for Britain?," New Scientist, 9th Feb. 1967, p. 346.

   e) On The Shelf - A Survey of Industrial R & D projects abandoned for non-technical reasons, Centre for the Study of Industrial Innovation, London, July '71.

   f) Prof. E.B. Roberts: 'Entrepreneurship & Technology,' in Gruber & Marguis (ed.): Factors in the Transfer of Technology, MIT Press, 1969.

   g) Gaps in Technology: Scientific Instruments, OECD,

3. Lorsch, Jay W. & Paul R. Lawrence, "Organizing for Product Innovation," Harvard Business Review (Jan.-Feb. 1965). They stress that the organization must provide a means so that the full energy of research, sales and production people can be brought to bear on innovation.

4. Jacob Schmookler, "Invention and Economic Growth" (Cambridge: Harvard Univ. Press; 1966). Schmookler argues that innovation may be induced by a combination of the two forces. The two forces being advances in knowledge called technology-push and rising demand as demand-pull innovations.

5. S. Myers and D.G. Marquis, Successful Commercial Innovations NSF 69-71 (NSF, Wash. D. C. '69); J. Langrish, R & D Management 1,133 (June 1971); J.M. Utterback, thesis, M.I.T. '69.

6. Marquis, D. Innovation #7, 1969.

7. J. Jewkes, D. Sawers, and R. Stillerman, "The Source of Invention," St. Martin's Press, N.Y. 1958; the study claims that the bulk of R&D carried out by large corporations is relatively short term.

8. Simon, Herbert A., "Theories of Decision-Making in Economic and Behavioral Science," American Economic Review, 1959; Simon points out that maximization of profits as a theory of the firm leaves ambiguous whether it is short-run or long-run profit that is to be examined.

9. Jay W. Forrester, "Advertising A Problem in Industrial Dynamics," Harvard Business Review, March-April 1959, p. 108. Forrester examines the four periods of a product life cycle-product introduction, market growth, market maturity and sales decline.

10. Dr. Christian Sommer, "Superheated Steam Fixation of Dyes in Printing on Synthetic Fibres," BASF, Ludwigshafen, Germany.

11. Samuelson, Paul A. "Economics," McGraw-Hill, N.Y. 1970; ". . . we must recognize that any invention which lowers cost of production can benefit the first competitor who introduces it."

12. Patent considerations means that the respondent answered "yes" to questions such as: Was patented technology used? Were licensing agreements necessary? Would a patent cover the innovative idea? This is consistent with the Chapter 3 findings which indicated - in a separate study of 1200 patents - that the large companies will often not embark upon innovation projects without patent protection.

NOTES

CHAPTER 7

1. Nasbeth & Ray stress the importance of information flow on innovation. Nasbeth, L. and Ray, G.F. (eds.). <u>The Diffusion on New Industrial Process: An International Study</u>. Cambridge University Press, 1974.

2. Lorsch, Jay W. & Paul R. Lawrence, "Organizing for Product Innovation," <u>Harvard Business Review</u> (Jan.-Feb. 1965). Lorsch & Lawrence stress the importance for managers interested in improving their record with new products to have effective means of internal coordination. They state that the organization must provide a means so that the full energy of research, sales, and production people can be brought to bear on innovation.

3. House Subcommittee on Science, Research and Development, "Technology Assessment," 1968; the concept of information was strongly emphasized.

4. The study of Schmitz analyzed some obstacles to diffusion. One of the conclusions was that a more frequent and efficient dialogue between allied industries was needed. Schmitz, N., "Obstacles a la Diffusion des Innovations a Belgique," Dulbea-Brussels, 1970. A study done in England showed that successful innovators make more effective use of outside technological advice and have better contacts with the outside community. Freeman, C. in William, B.R. (ed.) <u>Science and Technology in Economic Growth</u>, MacMillan, 1971; Robertson, A. "The Sappho Project: Success and Failure in the Innovation Process." Science Policy Research Unit, University of Sussex, Brighton, 1971.

5. One of many examples in the U.S. was the close working relationship between Kodak and Sylvania for the development of a highly reliable flash cube for use with the instamatic camera.

6. Allen, T.J., "Communications in R&D Organizations," Technology Review, 1967. Many of his studies have emphasized the importance of person-to-person information transfer.

NOTES

CHAPTER 8

1. Mansfield, Edwin, "The Science-Based Firm and the American Economy," from Micro-economics (Mansfield) Norton, N.Y. 1971; Mansfield stresses that the future role played by the science-based firm will be heavily dependent of the policies adopted by the government.

2. Helmar Krupp, (Director, Institute for Systems Engineering and Innovation, Karlsruhe, West Germany) made this point particularly during a talk in Brussels on November 7, 1974.

3. House Subcommittee on Science, Research, and Development, "Technology Assessment," 1968, The bill recognized the need for "identifying the potentials of applied research and technology and promoting ways and means to accomplish their transfer into practical use." One could similarly argue that the "radars" and antennas of society should be tuned to results of research and development so that applications become more frequent.

4. T.J. Allen, "The Diffusion of Technological Information," Proc. 20th Natl. Conference on the Administration of Research (Denver: Denver Research Inst., 1967); Allen shows that information flows within laboratories most readily through local opinion leaders.

T
175.5
G48
1977

MAR 7 1978